The Pleasure
of Finding Things Out

other books
by Richard P. Feynman

The Character of Physical Law

Elementary Particles and the Laws of Physics:
The 1986 Dirac Memorial Lectures (with Steven Weinberg)

Feynman Lectures on Physics (with Robert Leighton and Matthew Sands)

Lectures on Computation (edited by Anthony J. G. Hey and Robin Allen)

Lectures on Gravitation (with Fernando B. Morinigo and
William G. Wagner; edited by Brian Hatfield)

The Meaning of It All: Thoughts of a Citizen-Scientist

Photon-Hadron Interactions

QED: The Strange Theory of Light and Matter

Quantum Electrodynamics

Quantum Mechanics and Path Integrals (with A. R. Hibbs)

Six Easy Pieces: Essentials of Physics Explained by Its Most Brilliant Teacher

Six Not-So-Easy Pieces: Einstein's Relativity, Symmetry, and Space-Time

Statistical Mechanics: A Set of Lectures

Surely You're Joking, Mr. Feynman! Adventures of a Curious Character

The Theory of Fundamental Processes

What Do You Care What Other People Think?
Further Adventures of a Curious Character

The Pleasure
of Finding Things Out

*The Best Short Works
of Richard P. Feynman*

by Richard P. Feynman

*Edited by Jeffrey Robbins
Foreword by Freeman Dyson*

HELIX BOOKS

PERSEUS BOOKS
Cambridge, Massachusetts

Library of Congress Catalog Card Number: 99-64775
ISBN: 0-7382-0108-1

Perseus Books is a member of the Perseus Books Group

Jacket design Bruce W. Bond
Text design by Rachel Hegarty
Set in 11-point Berthold Garamond

1 2 3 4 5 6 7 8 9 10   03 02 01 00 99
First printing, July, 1999

Find Helix Books on the World Wide Web at
http://www.perseusbooks.com

Contents

Foreword:
This Side Idolatry
by Freeman Dyson

"I did love the man this side idolatry as much as any," wrote Elizabethan dramatist Ben Jonson. "The man" was Jonson's friend and mentor, William Shakespeare. Jonson and Shakespeare were both successful playwrights. Jonson was learned and scholarly, Shakespeare was slapdash and a genius. There was no jealousy between them. Shakespeare was nine years older, already filling the London stage with masterpieces before Jonson began to write. Shakespeare was, as Jonson said, "honest and of an open and free nature," and gave his young friend practical help as well as encouragement. The most important help that Shakespeare gave was to act one of the leading roles in Jonson's first play, "Every Man in His Humour," when it was performed in 1598. The play was a resounding success and launched Jonson's professional career. Jonson was then aged 25, Shakespeare 34. After 1598, Jonson continued to write poems and plays, and many of his plays were performed by Shakespeare's company. Jonson became famous in his own right as a poet and scholar, and at the end of his life he was honored with burial in Westminster Abbey. But he never forgot his debt to his old friend. When Shakespeare died, Jonson wrote a poem, "To the Memory of My Beloved

Master, William Shakespeare," containing the well-known lines:

"He was not of an age, but for all time."

"And though thou hadst small Latin and less Greek,
From thence to honor thee, I would not seek
For names, but call forth thundering Aeschylus,
Euripides and Sophocles, . . .
To live again, to hear thy buskin tread."

"Nature herself was proud of his designs,
And joyed to wear the dressing of his lines, . . .
Yet I must not give Nature all: Thy art,
My gentle Shakespeare, must enjoy a part.
For though the poet's matter nature be,
His art does give the fashion; and, that he
Who casts to write a living line, must sweat, . . .
For a good poet's made, as well as born."

What have Jonson and Shakespeare to do with Richard Feynman? Simply this. I can say as Jonson said, "I did love this man this side idolatry as much as any." Fate gave me the tremendous luck to have Feynman as a mentor. I was the learned and scholarly student who came from England to Cornell University in 1947 and was immediately entranced by the slapdash genius of Feynman. With the arrogance of youth, I decided that I could play Jonson to Feynman's Shakespeare. I had not expected to meet Shakespeare on American soil, but I had no difficulty in recognizing him when I saw him.

Before I met Feynman, I had published a number of mathematical papers, full of clever tricks but totally lacking in importance. When I met Feynman, I knew at once that I had entered another world. He was not interested in publishing pretty papers. He was struggling, more intensely than I had ever seen anyone struggle, to understand the workings of nature by rebuilding physics from the bottom up. I was lucky to meet him near the end of his eight-year struggle. The new physics that he had imagined as a student of John Wheeler seven years earlier was finally coalescing into a coherent vision of nature, the vision that he called "the space-time approach." The vision was in 1947 still unfinished, full of loose ends and inconsistencies, but I saw at once that it had to be right. I seized every opportunity to listen to Feynman talk, to learn to swim in the deluge of his ideas. He loved to talk, and he welcomed me as a listener. So we became friends for life.

For a year I watched as Feynman perfected his way of describing nature with pictures and diagrams, until he had tied down the loose ends and removed the inconsistencies. Then he began to calculate numbers, using his diagrams as a guide. With astonishing speed he was able to calculate physical quantities that could be compared directly with experiment. The experiments agreed with his numbers. In the summer of 1948 we could see Jonson's words coming true: "Nature herself was proud of his designs, and joyed to wear the dressing of his lines."

During the same year when I was walking and talking with Feynman, I was also studying the work of the physicists Schwinger and Tomonaga, who were following more conventional paths and arriving at similar results. Schwinger and Tomonaga had independently succeeded, using more laborious and complicated methods, in calculating the same quantities that Feynman could derive directly from his diagrams. Schwinger and Tomonaga did not rebuild physics. They took physics as they found it, and only introduced new mathematical methods to extract numbers from the physics. When it became clear that the results of their calculations agreed with Feynman, I knew that I had been given a unique opportunity to bring the three theories together. I wrote a paper with the title "The Radiation Theories of Tomonaga, Schwinger and Feynman," explaining why the theories looked different but were fundamentally the same. My paper was published in the *Physical Review* in 1949, and launched my professional career as decisively as "Every Man in His Humour" launched Jonson's. I was then, like Jonson, 25 years old. Feynman was 31, three years younger than Shakespeare had been in 1598. I was careful to treat my three protagonists with equal dignity and respect, but I knew in my heart that Feynman was the greatest of the three and that the main purpose of my paper was to make his revolutionary ideas accessible to physicists around the world. Feynman actively encouraged me to publish his ideas, and never once complained that I was stealing his thunder. He was the chief actor in my play.

One of the treasured possessions that I brought from England to America was "The Essential Shakespeare" by J. Dover Wilson, a short biography of Shakespeare containing most of the quotations from Jonson that I have reproduced here. Wilson's book is neither a work of fiction nor a work of history, but something in between. It is based on the first-hand testimony of Jonson and others, but Wilson used his imagination together with the scanty historical documents to bring Shakespeare to life. In particular, the earliest evidence that Shakespeare acted in Jonson's play comes from a document dated 1709, more than a hundred years after the event. We know that Shakespeare was famous as an actor as well as a writer, and I see no reason to doubt the traditional story as Wilson tells it.

Luckily, the documents that provide evidence of Feynman's life and thoughts are not so scanty. The present volume is a collection of such documents, giving us the authentic voice of Feynman recorded in his lectures and occasional writings. These documents are informal, addressed to general audiences rather than to his scientific colleagues. In them we see Feynman as he was, always playing with ideas but always serious about the things that mattered to him. The things that mattered were honesty, independence, willingness to admit ignorance. He detested hierarchy and enjoyed the friendship of people in all walks of life. He was, like Shakespeare, an actor with a talent for comedy.

Besides his transcendent passion for science, Feynman had also a robust appetite for jokes and ordinary human pleasures. A week after I got to know him, I wrote a letter to my parents in England describing him as "half genius and half buffoon." Between his heroic struggles to understand the laws of nature, he loved to relax with friends, to play his bongo drums, to entertain everybody with tricks and stories. In this too he resembled Shakespeare. Out of Wilson's book I take the testimony of Jonson:

"When he hath set himself to writing, he would join night to day; press upon himself without release, not minding it till he fainted: and when he left off, remove himself into all sports and looseness again; that it was almost a despair to draw him to his book: but once got to it, he grew stronger and more earnest by the ease."

That was Shakespeare, and that was also the Feynman I knew and loved, this side idolatry.

Freeman J. Dyson
Institute for Advanced Study
Princeton, New Jersey

Editor's Introduction

✦

Recently I was present at a lecture at Harvard University's venerable Jefferson Lab. The speaker was Dr. Lene Hau of the Rowland Institute, who had just conducted an experiment that was reported not only in the distinguished scientific journal *Nature* but also on the front page of the *New York Times*. In the experiment, she (with her research group of students and scientists) passed a laser beam through a new kind of matter called a Bose-Einstein condensate (a weird quantum state in which a bunch of atoms, cooled almost to absolute zero, practically stop moving at all and together act like a single particle), which slowed that light beam to the unbelievably leisurely pace of 38 miles per hour. Now light, which normally travels at the breakneck pace of 186,000 miles per *second*, or 669,600,000 miles per hour, in a vacuum, does typically slow down whenever it passes through any medium, such as air or glass, but only by a fraction of a percent of its speed in vacuo. But do the arithmetic and you will see that 38 miles per hour divided by 669.6 million miles per hour equals 0.00000006, or *six-millionths of a percent*, of its speed in vacuo. To put this result in perspective, it is as if Galileo had dropped his cannonballs from the Tower of Pisa and they took two years to reach the ground.

I was left breathless by the lecture (even Einstein would have been impressed, I think). For the first time in my life I felt a smidgen of what Richard Feynman called "the kick in the discovery," the sudden feeling (probably akin to an epiphany, albeit in this case a vicarious one) that I had grasped a wonderful new idea, that there was something new in the world; that I was present at a momentous scientific event, no less dramatic or exciting than Newton's feeling when he realized that the mysterious force that caused that apocryphal apple to land on his head was the same force that caused the moon to orbit the earth; or Feynman's when he achieved that first grudging step toward understanding the nature of the interaction between light and matter, which led eventually to his Nobel Prize.

Sitting among that audience, I could almost feel Feynman looking over my shoulder and whispering in my ear, "You see? That's why scientists persist in their investigations, why we struggle so desperately for every bit of knowledge, stay up nights seeking the answer to a problem, climb the steepest obstacles to the next fragment of understanding, to finally reach that joyous moment of the kick in the discovery, which is part of the pleasure of finding things out."* Feynman always said that he did physics not for the glory or for awards and prizes but for the *fun* of it, for the sheer pleasure of finding out how the world works, what makes it tick.

Feynman's legacy is his immersion in, and dedication to, science—its logic, its methods, its rejection of dogma, its infinite capacity to doubt. Feynman believed and lived by the credo that science, when used responsibly, can not only be fun but can also be of inestimable value to the future of

---

*Another of the most exciting events, if not in my life, then at least in my publishing career, was finding the long-buried, never-before-published transcript of three lectures Feynman gave at the University of Washington in the early 1960s, which became the book *The Meaning of It All*; but that was more the pleasure of finding things than the pleasure of finding things *out*.

human society. And like all great scientists, Feynman loved sharing his wonder of nature's laws with colleagues and laypersons alike. Nowhere is Feynman's passion for knowledge more clearly displayed than in this collection of his short works (most previously published, one unpublished).

The best way to appreciate the Feynman mystique is to read this book, for here you will find a wide range of topics about which Feynman thought deeply and discoursed so charmingly, not only physics—in the teaching of which he was surpassed by no one—but also religion, philosophy, and academic stage fright; the future of computing, and of nanotechnology, of which he was the first pioneer; humility, fun in science, and the future of science and civilization; how budding scientists should view the world; and the tragic bureaucratic blindness that led to the Space Shuttle *Challenger* disaster, the headline-making report that made "Feynman" a household word.

Remarkably, there is very little overlap in these pieces, but in those few places where a story is repeated in another piece, I took the liberty of deleting one of the two occurrences to spare the reader needless repetition. I inserted ellipses (...) to indicate where a repeated "gem" has been deleted.

Feynman had a very casual attitude toward proper grammar, as clearly shows in most of the pieces, which were transcribed from spoken lectures or interviews. To maintain the Feynman flavor, therefore, I generally let stand his ungrammatical turns of phrase. However, where poor or sporadic transcription made a word or phrase incomprehensible or awkward, I edited it for readability. I believe that the result is virtually unspoiled, yet readable, Feynmanese.

Acclaimed during his lifetime, revered in memory, Feynman continues to be a source of wisdom to people from all walks of life. I hope this treasury of his best talks, interviews,

and articles will stimulate and entertain generations of de-
voted fans and newcomers to Feynman's unique and often
rambunctious mind.

So read, enjoy, and don't be afraid to laugh out loud occasionally or to learn a lesson or two about life; be inspired;
above all, experience the pleasure of finding things out about
an uncommon human being.

I would like to thank Michelle and Carl Feynman for their
generosity and constant support from both coasts; Dr. Judith
Goodstein, Bonnie Ludt, and Shelley Erwin of the Caltech
archives for their indispensable help and hospitality; and especially professor Freeman Dyson for his elegant and en-
lightening Foreword.

I would also like to express my thanks to John Gribbin,
Tony Hey, Melanie Jackson, and Ralph Leighton for their frequent and excellent advice throughout the making of this
book.

Jeffrey Robbins,
Reading, Massachusetts,
September 1999

The Pleasure
of Finding Things Out

The Best Short Works of
Richard P. Feynman

1

The Pleasure of Finding Things Out

✦

This is the edited transcript of an interview with Feynman made for the BBC television program Horizon *in 1981, shown in the United States as an episode of* Nova. *Feynman had most of his life behind him by this time (he died in 1988), so he could reflect on his experiences and accomplishments with the perspective not often attainable by a younger person. The result is a candid, relaxed, and very personal discussion on many topics close to Feynman's heart: why knowing merely the name of something is the same as not knowing anything at all about it; how he and his fellow atomic scientists of the Manhattan Project could drink and revel in the success of the terrible weapon they had created while on the other side of the world in Hiroshima thousands of their fellow human beings were dead or dying from it; and why Feynman could just as well have gotten along without a Nobel Prize.*

The Beauty of a Flower

I have a friend who's an artist and he's sometimes taken a view which I don't agree with very well. He'll hold up a flower and say, "Look how beautiful it is," and I'll agree, I think. And he says—"you see, I as an artist can see how beautiful this is, but you as a scientist, oh, take this all apart and it becomes a dull thing." And I think that he's kind of nutty. First of all, the beauty that he sees is available to other people and to me, too, I believe, although I might not be quite as refined aesthetically as he is; but I can appreciate the beauty of a flower. At the same time I see much more about the flower than he sees. I can imagine the cells in there, the complicated actions inside which also have a beauty. I mean it's not just beauty at this dimension of one centimeter, there is also beauty at a smaller dimension, the inner structure. Also the processes, the fact that the colors in the flower evolved in order to attract insects to pollinate it is interesting—it means that insects can see the color. It adds a question: Does this aesthetic sense also exist in the lower forms? Why is it aesthetic? All kinds of interesting questions which shows that a science knowledge only adds to the excitement and mystery and the awe of a flower. It only adds; I don't understand how it subtracts.

Avoiding Humanities

I've always been very one-sided about science and when I was younger I concentrated almost all my effort on it. I didn't have time to learn and I didn't have much patience with what's called the humanities, even though in the university there were humanities that you had to take. I tried my best to avoid somehow learning anything and working at it. It was only afterwards, when I got older, that I got more relaxed, that I've spread out a little bit. I've learned to draw and I read a little bit, but I'm really still a very one-sided person and I don't know a great deal. I have a limited intelligence and I use it in a particular direction.

Tyrannosaurus in the Window

We had the *Encyclopaedia Britannica* at home and even when I was a small boy [my father] used to sit me on his lap and read to me from the *Encyclopaedia Britannica*, and we would read, say, about dinosaurs and maybe it would be talking about the brontosaurus or something, or the tyrannosaurus rex, and it would say something like, "This thing is twenty-five feet high and the head is six feet across," you see, and so he'd stop all this and say, "Let's see what that means. That would mean that if he stood in our front yard he would be high enough to put his head through the window but not quite because the head is a little bit too wide and it would break the window as it came by."

Everything we'd read would be translated as best we could into some reality and so I learned to do that—everything that I read I try to figure out what it really means, what it's really saying by translating and so (LAUGHS) I used to read the *Encyclopaedia* when I was a boy but with translation, you see, so it was very exciting and interesting to think there were animals of such magnitude—I wasn't frightened that there would be one coming in my window as a consequence of this, I don't think, but I thought that it was very, very interesting, that they all died out and at that time nobody knew why.

We used to go to the Catskill Mountains. We lived in New York and the Catskill Mountains was the place where people went in the summer; and the fathers—there was a big group of people there but the fathers would all go back to New York to work during the week and only come back on the weekends. When my father came he would take me for walks in the woods and tell me various interesting things that were going on in the woods—which I'll explain in a minute—but the other mothers seeing this, of course, thought this was wonderful and that the other fathers should take their sons for walks, and they tried to work on them but they didn't get anywhere at first and they wanted my father to take all the kids, but he didn't want to because he had a special relationship with me—we had a personal thing together—so it ended up that the other fathers had to take their children for walks the next weekend, and the next Monday when they were all back to work, all the kids were playing in the field and one kid said to me, "See that bird, what kind of a bird is that?" And I said, "I haven't the slightest idea what kind of a bird it is." He says, "It's a brown throated thrush," or something, "Your father doesn't tell you anything." But it was the opposite: my father *had* taught me. Looking at a bird he says, "Do you know what that bird is? It's a brown throated thrush; but in Portuguese it's a . . . in Italian a . . . ," he says "in Chinese it's a . . . , in Japanese a . . . ," etcetera. "Now," he says, "you know in all the languages you want to know what the name of that bird is and when you've finished with all that," he says, "you'll know absolutely nothing whatever about the bird. You only know about humans in different places and what they call the bird. Now," he says, "let's look at the bird."

He had taught me to notice things and one day when I was playing with what we call an express wagon, which is a little wagon which has a railing around it for children to play with that they can pull around. It had a ball in it—I remember this—it had a ball in it, and I pulled the wagon and I noticed something about the way the ball moved, so I went to my father and I said, "Say, Pop, I noticed something: When I pull the wagon the ball rolls to the back of the wagon, and when I'm pulling it along and I suddenly stop, the ball rolls to the front of the wagon," and I says, "why is that?" And he said, "That nobody knows," he said. "The general principle is that things that are moving try to keep on moving and things that are standing still tend to stand still unless you push on them hard." And he says, "This tendency is called inertia but nobody knows why it's true." Now that's a deep understanding— he doesn't give me a name, he knew the difference between knowing the name of something and knowing something, which I learnt very early. He went on to say, "If you look close you'll find the ball does not rush to the back of the wagon, but it's the back of the wagon that you're pulling against the ball; that the ball stands still or as a matter of fact from the friction starts to move forward really and doesn't move back." So I ran back to the little wagon and set the ball up again and pulled the wagon from under it and looking sideways and seeing indeed he was right—the ball never moved backwards in the wagon when I pulled the wagon forward. It moved backward relative to the wagon, but relative to the sidewalk it was moved forward a little bit, it's just [that] the wagon caught up with it. So that's the way I was educated by my father, with those kinds of examples and discussions, no pressure, just lovely interesting discussions.

Algebra for the Practical Man

My cousin, at that time, who was three years older, was in high school and was having considerable difficulty with his algebra and had a tutor come, and I was allowed to sit in a corner while (LAUGHS) the tutor would try to teach my cousin algebra, problems like $2x$ plus something. I said to my cousin then, "What're you trying to do?" You know, I hear him talking about x. He says, "What do *you* know—$2x + 7$ is equal to 15," he says "and you're trying to find out what x is." I says, "You mean 4." He says, "Yeah, but you did it with arithmetic, you have to do it by algebra," and that's why my cousin was never able to do algebra, because he didn't understand how he was supposed to do it. There was no way. I learnt algebra fortunately by not going to school and knowing the whole idea was to find out what x was and it didn't make any difference how you did it—there's no such thing as, you know, you do it by arithmetic, you do it by algebra—that was a false thing that they had invented in school so that the children who have to study algebra can all pass it. They had invented a set of rules which if you followed them without thinking could produce the answer: subtract 7 from both sides, if you have a multiplier divide both sides by the multiplier and so on, and a series of steps by which you could get the answer if you didn't understand what you were trying to do.

There was a series of math books, which started *Arithmetic for the Practical Man*, and then *Algebra for the Practical Man*, and then *Trigonometry for the Practical Man*, and I learned trigonometry for the practical man from that. I soon forgot it again because I didn't understand it very well but the series was coming out, and the library was going to get *Calculus for the Practical Man* and I knew by this time by reading the *Encyclopaedia* that calculus was an important subject and it was an interesting one and I ought to learn it. I was older now, I was perhaps thirteen; and then the calculus book finally came out and I was so excited and I went to the library to take it out and she looks at me and she says, "Oh, you're just a child, what are you taking this book out for, this book is a [book for adults]." So this was one of the few times in my life I was uncomfortable and I lied and I said it was for my father, he selected it. So I took it home and I learnt calculus from it and I tried to explain it to my father and he'd start to read the beginning of it and he found it confusing and it really bothered me a little bit. I didn't know that he was so limited, you know, that he didn't understand, and I thought it was relatively simple and straightforward and he didn't understand it. So that was the first time I knew I had learnt more in some sense than he.

Epaulettes and the Pope

One of the things that my father taught me besides physics (LAUGHS), whether it's correct or not, was a disrespect for respectable . . . for certain kinds of things. For example, when I was a little boy, and a rotogravure—that's printed pictures in newspapers—first came out in the *New York Times,* he used to sit me again on his knee and he'd open a picture, and there was a picture of the Pope and everybody bowing in front of him. And he'd say, "Now look at these humans. Here is one human standing here, and all these others are bowing. Now what is the difference? This one is the Pope"—he hated the Pope anyway—and he'd say, "the difference is epaulettes"—of course not in the case of the Pope, but if he was a general—it was always the uniform, the position, "but this man has the same human problems, he eats dinner like anybody else, he goes to the bathroom, he has the same kind of problems as everybody, he's a human being. Why are they all bowing to him? Only because of his name and his position, because of his uniform, not because of something special he did, or his honor, or something like that." He, by the way, was in the uniform business, so he knew what the difference was between the man with the uniform off and the uniform on; it's the same man for him.

He was happy with me, I believe. Once, though, when I came back from MIT–I'd been there a few years–he said to me, "Now," he said, "you've become educated about these things and there's one question I've always had that I've never understood very well and I'd like to ask you, now that you've studied this, to explain it to me," and I asked him what it was. And he said that he understood that when an atom made a transition from one state to another it emits a particle of light called a photon. I said, "That's right." And he says, "Well, now, is the photon in the atom ahead of time that it comes out, or is there no photon in it to start with?" I says, "There's no photon in, it's just that when the electron makes a transition it comes" and he says "Well, where does it come from then, how does it come out?" So I couldn't just say, "The view is that photon numbers aren't conserved, they're just created by the motion of the electron." I couldn't try to explain to him something like: the sound that I'm making now wasn't in me. It's not like my little boy who when he started to talk, suddenly said that he could no longer say a certain word–the word was "cat"–because his word bag has run out of the word cat (LAUGHS). So there's no word bag that you have inside so that you use up the words as they come out, you just make them as they go along, and in the same sense there was no photon bag in an atom and when the photons come out they didn't come from somewhere, but I couldn't do much better. He was not satisfied with me in the respect that I never was able to explain any of the things that he didn't understand (LAUGHS). So he was unsuccessful, he sent me through all these universities in order to find out these things and he never did find out (LAUGHS).

Invitation to the Bomb

[While working on his PhD thesis, Feynman was asked to join the project to develop the atomic bomb.] It was a completely different kind of a thing. It would mean that I would have to stop the research in what I was doing, which is my life's desire, to take time off to do this, which I felt I should do in order to protect civilization. Okay? So that was what I had to debate with myself. My first reaction was, well, I didn't want to get interrupted in my normal work to do this odd job. There was also the problem, of course, of any moral thing involving war. I wouldn't have much to do with that, but it kinda scared me when I realized what the weapon would be, and that since it might be possible, it must be possible. There was nothing that I knew that indicated that if we could do it they couldn't do it, and therefore it was very important to try to cooperate.

[In early 1943 Feynman joined Oppenheimer's team at Los Alamos.] With regard to moral questions, I do have something I would like to say about it. The original reason to start the project, which was that the Germans were a danger, started me off on a process of action which was to try to develop this first system at Princeton and then at Los Alamos, to try to make the bomb work. All kinds of attempts were made to redesign it to make it a worse bomb and so on. It was a project on which we all worked very, very hard, all co-operating together. And with any project like that you continue to work trying to get success, having decided to do it. But what I did—immorally I would say—was to not remember the reason that I said I was doing it, so that when the reason changed, because Germany was defeated, not the singlest thought came to my mind at all about that, that that meant

now that I have to reconsider why I am continuing to do this. I simply didn't think, okay?

Success and Suffering

[On 6 August 1945 the atomic bomb was exploded over Hiroshima.] The only reaction that I remember—perhaps I was blinded by my own reaction—was a very considerable elation and excitement, and there were parties and people got drunk and it would make a tremendously interesting contrast, what was going on in Los Alamos at the same time as what was going on in Hiroshima. I was involved with this happy thing and also drinking and drunk and playing drums sitting on the hood of—the bonnet of—a Jeep and playing drums with excitement running all over Los Alamos at the same time as people were dying and struggling in Hiroshima.

I had a very strong reaction after the war of a peculiar nature—it may be from just the bomb itself and it may be for some other psychological reasons, I'd just lost my wife or something, but I remember being in New York with my mother in a restaurant, immediately after [Hiroshima], and thinking about New York, and I knew how big the bomb in Hiroshima was, how big an area it covered and so on, and I realized from where we were—I don't know, 59th Street—that to drop one on 34th Street, it would spread all the way out here and all these people would be killed and all the things would be killed and there wasn't only one bomb available, but it was easy to continue to make them, and therefore that things were sort of doomed because already it appeared to me—very early, earlier than to others who were more optimistic—that international relations and the way people were behaving were no different than they had ever been before and that it was just going to go on the same way as any other thing and I was sure that it was going, therefore, to be used very soon. So I felt very un-comfortable and thought, really believed, that it was silly: I would see people building a bridge and I would say "they don't understand." I really believed that it was senseless to make anything because it would all be destroyed very soon anyway, but they didn't understand that and I had this very strange view of any construction that I would see, I would always think how foolish they are to try to make something. So I was really in a kind of depressive condition.

"I Don't Have to Be Good Because They Think I'm Going to Be Good."

[After the war Feynman joined Hans Bethe* at Cornell University. He turned down the offer of a job at Princeton's Institute for Advanced Study.] They [must have] expected me to be wonder-ful to offer me a job like this and I wasn't wonderful, and therefore I realized a new principle, which was that I'm not responsible for what other people think I am able to do; I don't have to be good because they think I'm going to be good. And somehow or other I could relax about this, and I thought to myself, I haven't done anything important and I'm never going to do anything important. But I used to enjoy physics and mathematical things and because I used to play with them it was in very short order [that I] worked the things out for which I later won the Nobel Prize.†

---

*(1906–) Winner of the 1967 Nobel Prize in Physics for contributions to the theory of nuclear reactions, especially for his discoveries concerning the energy production in stars. *Ed.*

†In 1965, the Nobel Prize for Physics was shared by Richard Feynman, Julian Schwinger, and Sin-Itiro Tomonaga for their fundamental work in quantum electrodynamics, and its deep consequences for the physics of el-ementary particles. *Ed.*

The Nobel Prize—Was It Worth It?

[Feynman was awarded a Nobel Prize for his work on quantum electrodynamics.] What I essentially did—and also it was done independently by two other people, [Sinitiro] Tomanaga in Japan and [Julian] Schwinger—was to figure out how to control, how to analyze and discuss the original quantum theory of electricity and magnetism that had been written in 1928; how to interpret it so as to avoid the infinities, to make calculations for which there were sensible results which have since turned out to be in exact agreement with every experiment which has been done so far, so that quantum electrodynamics fits experiment in every detail where it's applicable—not involving the nuclear forces, for instance—and it was the work that I did in 1947 to figure out how to do that, for which I won the Nobel Prize.

[BBC: *Was it worth the Nobel Prize?*] As a (LAUGHS) . . . I don't know anything about the Nobel Prize, I don't understand what it's all about or what's worth what, but if the people in the Swedish Academy decide that *x, y,* or *z* wins the Nobel Prize then so be it. I won't have anything to do with the Nobel Prize . . . it's a pain in the . . . (LAUGHS). I don't like honors. I appreciate it for the work that I did, and for people who appreciate it, and I know there's a lot of physicists who use my work, I don't need anything else, I don't think there's any sense to anything else. I don't see that it makes any point that someone in the Swedish Academy decides that this work is noble enough to receive a prize—I've already got the prize. The prize is the pleasure of finding the thing out, the kick in the discovery, the observation that other people use it [my work]—those are the real things, the honors are unreal to me. I don't believe in honors, it bothers me, honors bother, honors is epaulettes, honors is uniforms. My papa brought me up this way. I can't stand it, it hurts me.

When I was in high school, one of the first honors I got was to be a member of the Arista, which is a group of kids who got good grades–eh?–and everybody wanted to be a member of the Arista, and when I got into the Arista I discovered that what they did in their meetings was to sit around to discuss who else was worthy to join this wonderful group that we are–okay? So we sat around trying to decide who it was who would get to be allowed into this Arista. This kind of thing bothers me psychologically for one or another reason I don't understand myself–honors– and from that day to this [it] always bothered me. When I became a member of the National Academy of Sciences, I had ultimately to resign because that was another organization most of whose time was spent in choosing who was illustrious enough to join, to be allowed to join us in our organization, including such questions as [should] we physicists stick together because they've a very good chemist that they're trying to get in and we haven't got enough room for so-and-so. What's the matter with chemists? The whole thing was rotten because its purpose was mostly to decide who could have this honor–okay? I don't like honors.

The Rules of the Game

[From 1950 to 1988 Feynman was Professor of Theoretical Physics at the California Institute of Technology.] One way, that's kind of a fun analogy in trying to get some idea of what we're doing in trying to understand nature, is to imagine that the gods are playing some great game like chess, let's say, and you don't know the rules of the game, but you're allowed to look

at the board, at least from time to time, in a little corner, perhaps, and from these observations you try to figure out what the rules of the game are, what the rules of the pieces moving are. You might discover after a bit, for example, that when there's only one bishop around on the board that the bishop maintains its color. Later on you might discover the law for the bishop as it moves on the diagonal which would explain the law that you understood before—that it maintained its color— and that would be analagous to discovering one law and then later finding a deeper understanding of it. Then things can happen, everything's going good, you've got all the laws, it looks very good, and then all of a sudden some strange phenomenon occurs in some corner, so you begin to investigate that—it's castling, something you didn't expect. We're always, by the way, in fundamental physics, always trying to investigate those things in which we don't understand the conclusions. After we've checked them enough, we're okay.

The thing that doesn't fit is the thing that's the most interesting, the part that doesn't go according to what you expected. Also, we could have revolutions in physics: after you've noticed that the bishops maintain their color and they go along the diagonal and so on for such a long time and everybody knows that that's true, then you suddenly discover one day in some chess game that the bishop doesn't maintain its color, it changes its color. Only later do you discover a new possibility, that a bishop is captured and that a pawn went all the way down to the queen's end to produce a new bishop—that can happen but you didn't know it, and so it's very analagous to the way our laws are: They sometimes look positive, they keep on working and all of a sudden some little gimmick shows that they're wrong and then we have to investigate the conditions under which this bishop change of color happened and so forth, and gradually learn the new rule that explains it more deeply. Unlike the chess game, though, in [which] the rules become more complicated as you go along, in physics, when you discover new things, it looks more simple. It appears on the whole to be more complicated because we learn about a greater experience—that is, we learn about more particles and new things—and so the laws look complicated again. But if you realize all the time what's kind of wonderful—that is, if we expand our experience into wilder and wilder regions of experience—every once in a while we have these integrations when everything's pulled together into a unification, in which it turns out to be simpler than it looked before.

If you are interested in the ultimate character of the physical world, or the complete world, and at the present time our only way to understand that is through a mathematical type of reasoning, then I don't think a person can fully appreciate, or in fact can appreciate much of, these particular aspects of the world, the great depth of character of the universality of the laws, the relationships of things, without an understanding of mathematics. I don't know any other way to do it, we don't know any other way to describe it accurately . . . or to see the interrelationships without it. So I don't think a person who hasn't developed some mathematical sense is capable of fully appreciating this aspect of the world—don't misunderstand me, there are many, many aspects of the world that mathematics is unnecessary for, such as love, which are very delightful and wonderful to appreciate and to feel awed and mysterious about; and I don't mean to say that the only thing in the world is physics, but you were talking about physics and if that's what you're talking about, then to not know mathematics is a severe limitation in understanding the world.

Smashing Atoms

Well, what I'm working on in physics right now is a special problem which we've come up against and I'll describe what it is. You know that everything's made out of atoms, we've got that far already and most people know that already, and that the atom has a nucleus with electrons going around. The behavior of the electrons on the outside is now completely [known], the laws for it are well understood as far as we can tell in this quantum electrodynamics that I told you about. And after that was evolved, then the problem was how does the nucleus work, how do the particles interact, how do they hold together? One of the by-products was to discover fission and to make the bomb. But investigating the forces that hold the nuclear particles together was a long task. At first it was thought that it was an exchange of some sort of particles inside, which were invented by Yukawa, called pions, and it was predicted that if you hit protons—the proton is one of the particles of the nucleus—against a nucleus, they would knock out such pions, and sure enough, such particles came out.

Not only pions came out but other particles, and we began to run out of names—kaons and sigmas and lamdas and so on; they're all called hadrons now—and as we increased the energy of the reaction and got more and more different kinds, until there were hundreds of different kinds of particles; then the problem, of course—this period is 1940 up to 1950, towards the present—was to find the pattern behind it. There seemed to be many many interesting relations and patterns among the particles, until a theory was evolved to explain these patterns, that all of these particles were really made of something else, that they were made of things called quarks—three quarks, for example, would form a proton—and that the proton is one of the particles of the nucleus; another one is a neutron. The quarks came in a number of varieties—in fact, at first only three were needed to explain all the hundreds of particles and the different kinds of quarks—they are called u-type, d-type, s-type. Two us and a d made a proton, two ds and a u made a neutron. If they were moving in a different way inside they were some other particle. Then the problem came: What exactly is the behavior of the quarks and what holds them together? And a theory was thought of which is very simple, a very close analogy to quantum electrodynamics—not exactly the same but very close—in which the quarks are like the electron and the particles called gluons—which go between the electrons, which makes them attract each other electrically—are like the photons. The mathematics was very similar but there are a few terms slightly different. The difference in the form of the equations that were guessed at were guessed by principles of such beauty and simplicity that it isn't arbitrary, it's very, very determined. What is arbitrary is how many different kinds of quark there are, but not the character of the force between them.

Now unlike electrodynamics, in which two electrons can be pulled apart as far as you want, in fact when they are very far away the force is weakened; if this were true for quarks you would have expected that when you hit things together hard enough the quarks would have come out. But instead of that, when you're doing an experiment with enough energy that quarks could come out, instead of that you find a big jet—that is, all particles going about in the same direction as the old hadrons, no quarks—and from the theory, it was clear that what was required was that when the quark comes out, it kind of makes these new pairs of quarks and they come in little groups and make hadrons.

The question is, why is it so different in electrodynamics, how do these small-term differences, these little terms that are different in the equation, produce such different effects, entirely different effects? In fact, it was very surprising to most people that this would really come out, that first you would think that the theory was wrong, but the more it's studied the clearer it became that it's very possible that these extra terms would produce these effects. Now we were in a position that's different in history than any other time in physics, that's always different. We have a theory, a complete and definite theory of all of these hadrons, and we have an enormous number of experiments and lots and lots of details, so why can't we test the theory right away to find out whether it's right or wrong? Because what we have to do is calculate the consequences of the theory. If this theory is right, what should happen, and has that happened? Well, this time the difficulty is in the first step. If the theory is right, what should happen is very hard to figure out. The mathematics needed to figure out what the consequences of this theory are have turned out to be, at the present time, insuperably difficult. At the present time—all right? And therefore it's obvious what my problem is—my problem is to try to develop a way of getting numbers out of this theory, to test it really carefully, not just qualitatively, to see if it might give the right result.

I spent a few years trying to invent mathematical things that would permit me to solve the equations, but I didn't get anywhere, and then I decided that in order to do that I must first understand more or less how the answer probably looks. It's hard to explain this very well, but I had to get a qualitative idea of how the phenomenon works before I could get a good quantitative idea. In other words, people didn't even understand roughly how it worked, and so I have been working most recently in the last year or two on understanding roughly how it works, not quantitatively yet, with the hope that in the future that rough understanding can be refined into a precise mathematical tool, way, or algorithm to get from the theory to the particles. You see, we're in a funny position: It's not that we're looking for the theory, we've got the theory—a good, good candidate—but we're in the step in the science that we need to compare the theory to experiment by seeing what the consequences are and checking it. We're stuck in seeing what the consequences are, and it's my aim, it's my desire to see if I can work out a way to work out what the consequences of this theory are (LAUGHS). It's a kind of a crazy position to be in, to have a theory that you can't work out the consequences of . . . I can't stand it, I have to figure it out. Someday, maybe.

"Let George Do It."

To do high, real good physics work you do need absolutely solid lengths of time, so that when you're putting ideas together which are vague and hard to remember, it's very much like building a house of cards and each of the cards is shaky, and if you forget one of them the whole thing collapses again. You don't know how you got there and you have to build them up again, and if you're interrupted and kind of forget half the idea of how the cards went together—your cards being different-type parts of the ideas, ideas of different kinds that have to go together to build up the idea—the main point is, you put the stuff together, it's quite a tower and it's easy [for it] to slip, it needs a lot of concentration—that is, solid time to think—and if you've got a job in administrating anything like that, then you don't have the solid time. So I have invented another myth for myself—that I'm irresponsible. I tell everybody, I don't do anything. If anybody asks me to be on a committee to take care of admissions, no, I'm irresponsible, I don't give a damn about the students—of course I give a damn about the students but I know that somebody else'll do it—and I take the view, "Let George do it," a view which you're not supposed to take, okay, because that's not right to do, but I do that because I like to do physics and I want to see if I can still do it, and so I'm selfish, okay? I want to do my physics.

Bored by the History

All those students are in the class: Now you ask me how should I best teach them? Should I teach them from the point of view of the history of science, from the applications? My theory is that the best way to teach is to have no philosophy, [it] is to be chaotic and [to] confuse it in the sense that you use every possible way of doing it. That's the only way I can see to answer it, so as to catch this guy or that guy on different hooks as you go along, [so] that during the time when the fellow who's interested in history's being bored by the abstract mathematics, on the other hand the fellow who likes the abstractions is being bored another time by the history—if you can do it so you don't bore them all, all the time, perhaps you're better off. I really don't know how to do it. I don't know how to answer this question of different kinds of minds with different kinds of interests—what hooks them on, what makes them interested, how you direct them to become interested. One way is by a kind of force, you have to pass this course, you have to take this examination. It's a very effective way. Many people go through schools that way and it may be a more effective way. I'm sorry, after many, many years of trying to teach and trying all different kinds of methods, I really don't know how to do it.

Like Father, Like Son

I got a kick, when I was a boy, [out] of my father telling me things, so I tried to tell my son things that were interesting about the world. When he was very small we used to rock him to bed, you know, and tell him stories, and I'd make up a story about little people that were about so high [who] would walk along and they would go on picnics and so on and they lived in the ventilator; and they'd go through these woods which had great big long tall blue things like trees, but without leaves and only one stalk, and they had to walk between them and so on; and he'd gradually catch on [that] that was the rug, the nap of the rug, the blue rug, and he loved this game because I would describe all these things from an odd point of view and he liked to hear the stories and we got all kinds of wonderful things—he even went to a moist cave where the wind kept going in and out—it was coming in cool and went out warm and so on. It was inside the dog's nose that they went, and then of course I could tell him all about physiology by this way and so on. He loved that and so I told him lots of stuff, and I enjoyed it because I was telling him stuff that I liked, and we had fun when he would guess what it was and so on. And then I have a daughter and I tried the same thing—well, my daughter's personality was different, she didn't want to hear this story, she wanted the story that was in the book repeated again, and reread to her. She wanted me to read to her, not to make up stories, and it's a different personality. And so if I were to say a very good method for teaching children about science is to make up these stories of the little people, it doesn't work at all on my daughter—it happened to work on my son—okay?

"Science Which Is Not a Science . . . "

Because of the success of science, there is, I think, a kind of pseudoscience. Social science is an example of a science which is not a science; they don't do [things] scientifically, they follow the forms—or you gather data, you do so-and-so and so forth but they don't get any laws, they haven't found out anything. They haven't got anywhere yet—maybe someday they will, but it's not very well developed, but what happens is on an even more mundane level. We get experts on everything that sound like they're sort of scientific experts. They're not scientific, they sit at a typewriter and they make up something like, oh, food grown with, er, fertilizer that's organic is better for you than food grown with fertilizer that's inorganic—may be true, may not be true, but it hasn't been demonstrated one way or the other. But they'll sit there on the typewriter and make up all this stuff as if it's science and then become an expert on foods, organic foods and so on. There's all kinds of myths and pseudoscience all over the place.

I may be quite wrong, maybe they do know all these things, but I don't think I'm wrong. You see, I have the advantage of having found out how hard it is to get to really know something, how careful you have to be about checking the experiments, how easy it is to make mistakes and fool yourself. I know what it means to know something, and therefore I see how they get their information and I can't believe that they know it, they haven't done the work necessary, haven't done the checks necessary, haven't done the care necessary. I have a great suspicion that they don't know, that this stuff is [wrong] and they're intimidating people. I think so. I don't know the world very well but that's what I think.

Doubt and Uncertainty

If you expected science to give all the answers to the wonderful questions about what we are, where we're going, what the meaning of the universe is and so on, then I think you could easily become disillusioned and then look for some mystic answer to these problems. How a scientist can take a mystic answer I don't know because the whole spirit is to understand—well, never mind that. Anyhow, I don't understand that, but anyhow if you think of it, the way I think of what we're doing is we're exploring, we're trying to find out as much as we can about the world. People say to me, "Are you looking for the ultimate laws of physics?" No, I'm not, I'm just looking to find out more about the world and if it turns out there is a simple ultimate law which explains everything, so be it, that would be very nice to discover.

If it turns out it's like an onion with millions of layers and we're just sick and tired of looking at the layers, then that's the way it is, but whatever way it comes out its nature is there and she's going to come out the way she is, and therefore when we go to investigate it we shouldn't predecide what it is we're trying to do except to try to find out more about it. If you say your problem is, why do you find out more about it, if you thought you were trying to find out more about it because you're going to get an answer to some deep philosophical question, you may be wrong. It may be that you can't get an answer to that particular question by finding out more about the character of nature, but I don't look at it [like that]. My interest in science is to simply find out about the world, and the more I find out the better it is, like, to find out.

The Pleasure of Finding Things Out

There are very remarkable mysteries about the fact that we're able to do so many more things than apparently animals can do, and other questions like that, but those are mysteries I want to investigate without knowing the answer to them, and so altogether I can't believe these special stories that have been made up about our relationship to the universe at large because they seem to be too simple, too connected, too local, too provincial. The earth, He came to the earth, one of the aspects of God came to the earth, mind you, and look at what's out there. It isn't in proportion. Anyway, it's no use arguing, I can't argue it, I'm just trying to tell you why the scientific views that I have do have some effect on my belief. And also another thing has to do with the question of how you find out if something's true, and if all the different religions have all different theories about the thing, then you begin to wonder. Once you start doubting, just like you're supposed to doubt, you ask me if the science is true. You say no, we don't know what's true, we're trying to find out and everything is possibly wrong.

Start out understanding religion by saying everything is possibly wrong. Let us see. As soon as you do that, you start sliding down an edge which is hard to recover from and so on. With the scientific view, or my father's view, that we should look to see what's true and what may be or may not be true, once you start doubting, which I think to me is a very fundamental part of my soul, to doubt and to ask, and when you doubt and ask it gets a little harder to believe.

You see, one thing is, I can live with doubt and uncertainty and not knowing. I think it's much more interesting to live not knowing than to have answers which might be wrong. I have approximate answers and possible beliefs and different degrees of certainty about different things, but I'm not absolutely sure of anything and there are many things I don't know anything about, such as whether it means anything to ask why we're here, and what the question might mean. I might think about it a little bit and if I can't figure it out, then I go on to something else, but I don't have to know an answer, I don't feel frightened by not knowing things, by being lost in a mysterious universe without having any purpose, which is the way it really is so far as I can tell. It doesn't frighten me.

2

Computing Machines in the Future

✦

Forty years to the day after the atomic bombing of Nagasaki, Man-hattan Project veteran Feynman delivers a talk in Japan, but the topic is a peaceful one, one that still occupies our sharpest minds: the future of the computing machine, including the topic that made Feynman seem a Nostradamus of computer science—the ultimate lower limit to the size of a computer. This chapter may be challenging for some read-ers; however, it is such an important part of Feynman's contribution to science that I hope they will take the time to read it, even if they have to skip over some of the more technical spots. It ends with a brief discussion of one of Feynman's favorite pet ideas, which launched the current revolution in nanotechnology.

Introduction

It's a great pleasure and an honor to be here as a speaker in memorial for a scientist that I have respected and admired as much as Professor Nishina. To come to Japan and talk about computers is like giving a sermon to Buddha. But I have been thinking about computers and this is the only subject I could think of when invited to talk.

The first thing I would like to say is what I am not going to talk about. I want to talk about the future of computing machines. But the most important possible developments in the future are things that I will not speak about. For example, there is a great deal of work to try to develop smarter machines, machines which have a better relationship with humans so that input and output can be made with less effort than the complex programming that's necessary today. This often goes under the name of artificial intelligence, but I don't like that name. Perhaps the unintelligent machines can do even better than the intelligent ones.

Another problem is the standardization of programming languages. There are too many languages today, and it would be a good idea to choose just one. (I hesitate to mention that in Japan, for what will happen will be that there will simply be more standard languages—you already have four ways of writing now, and attempts to standardize anything here result apparently in more standards and not fewer!)

Another interesting future problem that is worth working on but I will not talk about is automatic debugging programs. Debugging means fixing errors in a program or in a machine, and it is surprisingly difficult to debug programs as they get more complicated.

Another direction of improvement is to make physical machines three dimensional instead of all on a surface of a chip. That can be done in stages instead of all at once—you can have several layers and then add many more layers as time goes on. Another important device would be one that could automatically detect defective elements on a chip; then the chip would automatically rewire itself so as to avoid the defective elements. At the present time, when we try to make big chips there are often flaws or bad spots in the chips, and we throw the whole chip away. If we could make it so that we could use the part of the chip that was effective, it would be much more efficient. I mention these things to try to tell you that I am aware of what the real problems are for future machines. But what I want to talk about is simple, just some small technical, physically good things that can be done in principle according to the physical laws. In other words, I would like to discuss the machinery and not the way we use the machines.

I will talk about some technical possibilities for making machines. There will be three topics. One is parallel processing machines, which is something of the very near future, almost present, that is being developed now. Further in the future is the question of the energy consumption of machines, which seems at the moment to be a limitation, but really isn't. Finally I will talk about the size. It is always better to make the machines smaller, and the question is, how much smaller is it still possible, in principle, to make machines according to the laws of Nature? I will not discuss which and what of these things will actually appear in the future. That depends on economic problems and social problems and I am not going to try to guess at those.

Parallel Computers

The first topic concerns parallel computers. Almost all the present computers, conventional computers, work on a layout or an architecture invented by von Neumann,* in which there is a very large memory that stores all the information, and one central location that does simple calculations. We take a number from this place in the memory and a number

---

*John von Neumann (1903–1957), a Hungarian-American mathematician who is credited as being one of the fathers of the computer. *Ed.*

from that place in the memory, send the two to the central arithmetical unit to add them, and then send the answer to some other place in the memory. There is, therefore, effectively one central processor which is working very, very fast and very hard, while the whole memory sits out there like a fast filing cabinet of cards which are very rarely used. It is obvious that if there were more processors working at the same time we ought to be able to do calculations faster. But the problem is that someone who might be using one processor may be using some information from the memory that another one needs, and it gets very confusing. For such reasons it has been said that it is very difficult to get many processors to work in parallel.

Some steps in that direction have been taken in the larger conventional machines called "vector processors." When sometimes you want to do exactly the same step on many different items, you can perhaps do that at the same time. The hope is that regular programs can be written in the ordinary way, and then an interpreter program will discover automatically when it is useful to use this vector possibility. That idea is used in the Cray and in "supercomputers" in Japan. Another plan is to take what is effectively a large number of relatively simple (but not very simple) computers, and connect them all together in some pattern. Then they can all work on a part of the problem. Each one is really an independent computer, and they will transfer information to each other as one or another needs it. This kind of a scheme is realized in the Caltech Cosmic Cube, for example, and represents only one of many possibilities. Many people are now making such machines. Another plan is to distribute very large numbers of very simple central processors all over the memory. Each one deals with just a small part of the memory and there is an elaborate system of interconnections between them. An example of such a machine is the Connection Machine made at MIT. It has 64,000 processors and a system of routing in which every 16 can talk to any other 16 and thus has 4,000 routing connection possibilities.

It would appear that scientific problems such as the propagation of waves in some material might be very easily handled by parallel processing. This is because what happens in any given part of space at any moment can be worked out locally and only the pressures and the stresses from the neighboring volumes need to be known. These can be worked out at the same time for each volume and these boundary conditions communicated across the different volumes. That's why this type of design works for such problems. It has turned out that a very large number of problems of all kinds can be dealt with in parallel. As long as the problem is big enough so that a lot of calculating has to be done, it turns out that a parallel computation can speed up time to solution enormously, and this principle applies not just to scientific problems.

What happened to the prejudice of two years ago, which was that the parallel programming is difficult? It turns out that what was difficult, and almost impossible, is to take an ordinary program and automatically figure out how to use the parallel computation effectively on that program. Instead, one must start all over again with the problem, appreciating that we have the possibility of parallel calculation, and rewrite the program completely with a new [understanding of] what is inside the machine. It is not possible to effectively use the old programs. They must be rewritten. That is a great disadvantage to most industrial applications and has met with considerable resistance. But the big programs usually belong to scientists or other, unofficial, intelligent programmers who love computer science and are willing to start all over again and rewrite the program if they can make it more efficient. So what's going to happen is that the hard programs, vast big ones, will be the first to be re-programmed by experts in the new way, and then gradually everybody will have to come around, and more and more programs will be programmed that way, and programmers will just have to learn how to do it.

Reducing the Energy Loss

The second topic I want to talk about is energy loss in computers. The fact that they must be cooled is an apparent limitation for the largest computers—a good deal of effort is spent in cooling the machine. I would like to explain that this is simply a result of very poor engineering and is nothing fundamental at all. Inside the computer a bit of information is controlled by a wire which has a voltage of either one value or another value. It is called "one bit," and we have to change the voltage of the wire from one value to the other and put charge on or take charge off. I make an analogy with water: We have to fill a vessel with water to get one level or empty it to get to the other level. This is just an analogy—if you like electricity better you can think more accurately electrically. What we do now is analogous, in the water case, to filling the vessel by pouring water in from a top level (Fig. 1), and lowering the level by opening the valve at the bottom and letting it all run out. In both cases there is a loss of energy because of the sudden drop in level of the water, through a height from the top level where it comes in, to the low bottom level, and also when you start pouring water in to fill it up again. In the cases of voltage and charge, the same thing occurs.

It's like, as Mr. Bennett has explained, operating an automobile which has to start by turning on the engine and stop

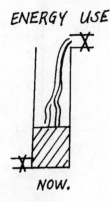

ENERGY USE

NOW.

FIGURE 1

by putting on the brakes. By turning on the engine and then putting on the brakes, each time you lose power. Another way to arrange things for a car would be to connect the wheels to flywheels. Now when the car stops, the flywheel speeds up, thus saving the energy—it can then be reconnected to start the car again. The water analog of this would be to have a U-shaped tube with a valve in the center at the bottom, connecting the two arms of the U (Fig. 2). We start with it full on the right but empty on the left with the valve closed. If we now open the valve, the water will slip over to the other side, and we can close the valve again, just in time to catch the water in the left arm. Now when we want to go the other way, we open the valve again and the water slips back to the other side and we catch it again. There is some loss and the water doesn't climb as high as it did before, but all we have to do is to put a little water in to correct the loss— a much smaller energy loss than the direct fill method. This trick uses the inertia of the water and the analog for electricity is inductance. However, it is very difficult with the silicon transistors that we use today to make up inductance on the

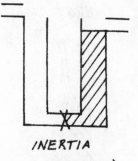

INERTIA

(INDUCTANCE)

FIGURE 2

chips. So this technique is not particularly practical with present technology.

Another way would be to fill the tank by a supply which stays only a little bit above the level of the water, lifting the water supply in time as we fill up the tank (Fig. 3), so that the dropping of water is always small during the entire effort. In the same way, we could use an outlet to lower the level in the tank, but just take water off near the top and lower the tube so that the heat loss would not appear at the position of the transistor, or would be small. The actual amount of loss will depend on how high the distance is between the supply and the surface as we fill it up. This method corresponds to changing the voltage supply with time. So if we could use a time varying voltage supply, we could use this method. Of course, there is energy loss in the voltage supply, but that is all located in one place and there it is simple to make one big inductance. This scheme is called "hot clocking," because the voltage supply operates at the same time as the clock which times everything. In addition, we don't need an extra clock signal to time the circuits as we do in conventional designs.

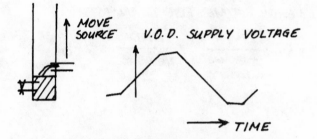

VARIABLE VOLTAGE SUPPLY
("HOT CLOCKING")
ENERGY LOSS · TIME = CONSTANT.

FIGURE 3

Both of these last two devices use less energy if they go slower. If I try to move the water supply level too fast, the water in the tube doesn't keep up with it and there ends being a big drop in water level. So to make the device work I must go slowly. Similarly, the U-tube scheme will not work unless that central valve can open and close faster than the time it takes for the water in the U-tube to slip back and forth. So my devices must be slower—I've saved an energy loss but I've made the devices slower. In fact the energy loss multiplied by the time it takes for the circuit to operate is constant. But nevertheless, this turns out to be very practical because the clock time is usually much larger than the circuit time for the transistors, and we can use that to decrease the energy. Also if we went, let us say, three times slower with our calculations, we could use one-third the energy over three times the time, which is nine times less power that has to be dissipated. Maybe this is worth it. Maybe by redesigning using parallel computations or other devices, we can spend a little longer than we could do at maximum circuit speed, in order to make

ENERGY · TIME FOR TRANSISTOR

$$= kT \cdot \frac{LENGTH}{\substack{THERMAL \\ VELOCITY}} \cdot \frac{LENGTH}{\substack{MEAN\ FREE \\ PATH}} \cdot \substack{NUMBER \\ OF \\ ELECTRONS}$$

ENERGY ~ 10^{9-11} kT

∴ DECREASE SIZE : FASTER
                   LESS ENERGY

FIGURE 4

a larger machine that is practical and from which we could still reduce the energy loss.

For a transistor, the energy loss multiplied by the time it takes to operate is a product of several factors (Fig. 4):

1. the thermal energy proportional to temperature, kT;
2. the length of the transistor between source and drain, divided by the velocity of the electrons inside (the thermal velocity $\sqrt{3kT/m}$);
3. the length of the transistor in units of the mean free path for collisions of electrons in the transistor;
4. the total number of the electrons that are inside the transistor when it operates.

Putting in appropriate values for all of these numbers tells us that the energy used in transistors today is somewhere between a billion to ten billion or more times the thermal energy kT. When the transistor switches, we use that much energy. This is very large amount of energy. It is obviously a good idea to decrease the size of the transistor. We decrease the length between source and drain and we can decrease the number of the electrons, and so use much less energy. It also turns out that a smaller transistor is much faster, because the electrons can cross it faster and make their decisions to switch faster. For every reason, it is a good idea to make the transistor smaller, and everybody is always trying to do that.

But suppose we come to a circumstance in which the mean free path is longer than the size of the transistor; then we discover that the transistor doesn't work properly anymore. It does not behave the way we expected. This reminds me, years ago there was something called the sound barrier. Airplanes were supposed not to be able to go faster than the speed of sound because, if you designed them normally and then tried to put the speed of sound in the equations, the propeller wouldn't work and the wings don't lift and nothing works correctly. Nevertheless, airplanes can go faster than the speed of sound. You just have to know what the right laws are under the right circumstances, and design the device with the correct laws. You cannot expect old designs to work in new circumstances. But *new* designs can work in *new* circumstances, and I assert that it is perfectly possible to make transistor systems, or, more correctly, switching systems and computing devices, in which the dimensions are smaller than the mean free path. I speak, of course, "in principle," and I am not speaking about the actual manufacture of such devices. Let us therefore discuss what happens if we try to make the devices as small as possible.

Reducing the Size

So my third topic is the size of computing elements and now I speak entirely theoretically. The first thing that you would

BROWNIAN MOTION

$$2 \ VOLT = 80 \ kT$$
$$PROB. \ ERROR \ e^{-80} = 10^{-43}$$

10^9 TRANSISTORS
10^{10} CHANGES / SEC. EACH
10^9 SECONDS (30 YEARS)

10^{28}

FIGURE 5

worry about when things get very small is Brownian motion*—everything is shaking about and nothing stays in place. How can you control the circuits then? Furthermore, if a circuit does work, doesn't it now have a chance of accidentally jumping back? If we use two volts for the energy of this electric system, which is what we ordinarily use (Fig. 5), that is eighty times the thermal energy at room temperature ($kT = $ 1/40 volt) and the chance that something jumps backward against 80 times thermal energy is e, the base of the natural logarithm, to the power minus eighty, or 10^{-43}. What does that mean? If we had a billion transistors in a computer (which we don't yet have), all of them switching 10^{10} times a second (a switching time of a tenth of a nanosecond), switching perpetually, operating for 10^9 seconds, which is 30 years, the total number of switching operations in such a machine is 10^{28}. The chance of one of the transistors going backward

---

*The jerky movements of particles caused by the constant random collisions of molecules, first noted in print in 1928 by botanist Robert Brown, and explained by Albert Einstein in a 1905 paper in *Annalen der Physik*. Ed.

is only 10^{-43}, so there will be no error produced by thermal oscillations whatsoever in 30 years. If you don't like that, use 2.5 volts and then the probability gets even smaller. Long before that, real failures will come when a cosmic ray accidentally goes through the transistor, and we don't have to be more perfect than that.

However, much more is in fact possible and I would like to refer you to an article in a most recent *Scientific American* by C. H. Bennett and R. Landauer, "The Fundamental Physical Limits of Computation."[*] It is possible to make a computer in which each element, each transistor, can go forward and accidentally reverse and still the computer will operate. All the operations in the computer can go forward or backward. The computation proceeds for a while one way and then it undoes itself, "uncalculates," and then goes forward again and so on. If we just pull it along a little, we can make this computer go through and finish the calculation by making it just a little bit more likely that it goes forward than backward.

It is known that all possible computations can be made by putting together some simple elements like transistors; or, if we want to be more logically abstract, something called a NAND gate, for example (NAND means NOT-AND). A NAND gate has two "wires" in and one out (Fig. 6). Forget the NOT for the moment. What is an AND gate? An AND gate is a device whose output is 1 only if both input wires are 1, otherwise its output is 0. NOT-AND means the opposite, thus the output wire reads 1 (i.e., has the voltage level corresponding to 1) unless both input wires read 1; if both input wires read 1, then the output wire reads 0 (i.e., has the voltage level corresponding to 0). Figure 6 shows a little table of inputs and outputs for such a NAND gate. *A* and *B* are in-

---

[*]*Sci. Am.* July 1985; Japanese Transl.–SAIENSU, Sept. 1985. *Ed.*

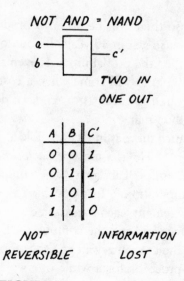

NOT AND = NAND

TWO IN
ONE OUT

A	B	C'
0	0	1
0	1	1
1	0	1
1	1	0

NOT   INFORMATION
REVERSIBLE   LOST

FIGURE 6

puts and C is the output. If A and B are both 1, the output is 0, otherwise 1. But such a device is irreversible: Information is lost. If I only know the output, I cannot recover the input. The device can't be expected to flip forward and then come back and compute correctly anymore. For instance, if we know that the output is now 1, we don't know whether it came from $A=0$, $B=1$ or $A=1$, $B=0$ or $A=0$, $B=0$ and it cannot go back. Such a device is an irreversible gate. The great discovery of Bennett and, independently, of Fredkin is that it is possible to do computation with a different kind of fundamental gate unit, namely, a reversible gate unit. I have illustrated their idea—with a unit which I could call a reversible NAND gate. It has three inputs and three outputs (Fig. 7). Of the outputs, two, A' and B', are the same as two of the inputs, A and B, but the third input works this way. C' is the same as C unless A and B are both 1, in which case it changes whatever C is. For instance, if C is 1 it is changed to 0, if C is 0 it

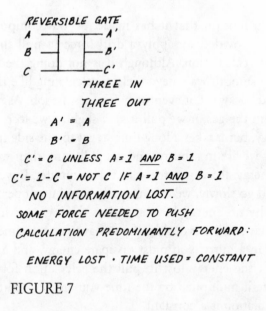

FIGURE 7

is changed to 1—but these changes only happen if both A and B are 1. If you put two of these gates in succession, you see that A and B will go through, and if C is not changed in both it stays the same. If C is changed, it is changed twice so that it stays the same. So this gate can reverse itself and no information has been lost. It is possible to discover what went in if you know what came out.

A device made entirely with such gates will make calculations if everything moves forward. But if things go back and forth for a while, but then eventually go forward enough, it still operates correctly. If the things flip back and then go forward later it is still all right. It's very much the same as a particle in a gas which is bombarded by the atoms around it. Such a particle usually goes nowhere, but with just a little pull, a little prejudice that makes a chance to move one way a little higher than the other way, the thing will slowly drift forward and travel from one end to the other, in spite of the

Brownian motion that it has made. So our computer will compute provided we apply a drift force to pull the thing across the calculation. Although it is not doing the calculation in a smooth way, nevertheless, calculating like this, forward and backward, it eventually finishes the job. As with the particle in the gas, if we pull it very slightly, we lose very little energy, but it takes a long time to get to one side from the other. If we are in a hurry, and we pull hard, then we lose a lot of energy. It is the same with this computer. If we are patient and go slowly, we can make the computer operate with practically no energy loss, even less than kT per step, any amount as small as you like if you have enough time. But if you are in a hurry, you must dissipate energy, and again it's true that the energy lost to pull the calculation forward to complete it multiplied by the time you are allowed to make the calculation is a constant.

With these possibilities in mind, let's see how small we can make a computer. How big must a number be? We all know we can write numbers in base 2 as strings of "bits," each a one or a zero. And the next atom could be a one or a zero, so a little string of atoms are enough to hold a number, one atom for each bit. (Actually, since an atom can have more than just two states, we could use even fewer atoms, but one per bit is little enough!) So, for intellectual entertainment, we consider whether we could make a computer in which the writing of bits is of atomic size, in which a bit is, for example, whether the spin in the atom is up for 1 or down for 0. And then our "transistor," which changes the bits in different places, would correspond to some interaction between atoms which will change their states. The simplest example would be if a kind of 3-atom interaction were to be the fundamental element or gate in such a computer. But again, the device won't work right if we design it with the laws appropriate for large objects.

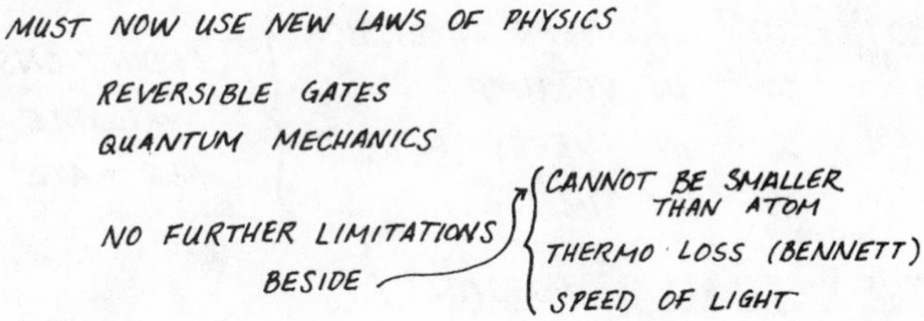

MUST NOW USE NEW LAWS OF PHYSICS

   REVERSIBLE GATES
   QUANTUM MECHANICS

                               { CANNOT BE SMALLER
                                        THAN ATOM
   NO FURTHER LIMITATIONS }{ THERMO LOSS (BENNETT)
          BESIDE ——————{ SPEED OF LIGHT

FIGURE 8

We must use the new laws of physics, quantum mechanical laws, the laws that are appropriate to atomic motion (Fig. 8).

We therefore have to ask whether the principles of quantum mechanics permit an arrangement of atoms so small in number as a few times the number of gates in a computer that could operate as a computer. This has been studied in principle, and such an arrangement has been found. Since the laws of quantum mechanics are reversible, we must use the invention by Bennett and Fredkin of reversible logic gates. When this quantum mechanical situation is studied, it is found that quantum mechanics adds no further limitations to anything that Mr. Bennett has said from thermodynamic considerations. Of course there is a limitation, the practical limitation anyway, that the bits must be of the size of an atom and a transistor 3 or 4 atoms. The quantum mechanical gate I used has 3 atoms. (I would not try to write my bits onto nuclei, I'll wait till the technological development reaches atoms before I need to go any further!) That leaves us just with: (a) the limitations in size to the size of atoms; (b) the energy requirements depending on the time as worked out by Bennett; and c) the feature that I did not mention concerning the speed of light—we can't send the signals any faster than the speed of light. These are the only physical limitations on computers that I know of.

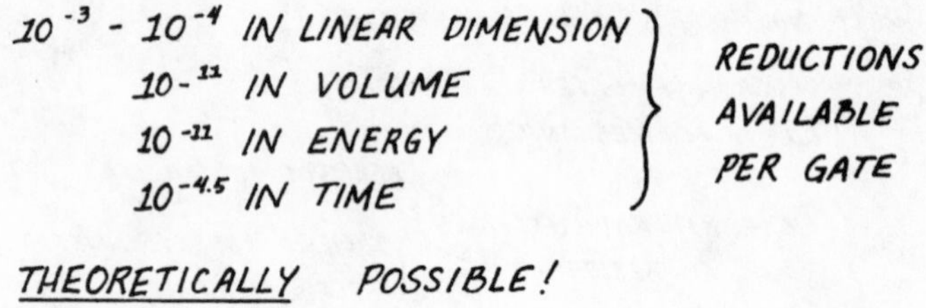

$$10^{-3} - 10^{-4} \text{ IN LINEAR DIMENSION}$$
$$10^{-11} \text{ IN VOLUME}$$
$$10^{-11} \text{ IN ENERGY}$$
$$10^{-4.5} \text{ IN TIME}$$

} REDUCTIONS AVAILABLE PER GATE

THEORETICALLY POSSIBLE!

FIGURE 9

If we somehow manage to make an atomic size computer, it would mean (Fig. 9) that the dimension, the *linear* dimension, is a thousand to ten thousand times smaller than those very tiny chips that we have now. It means that the volume of the computer is 100 billionth or 10^{-11} of the present volume, because the volume of the "transistor" is smaller by a factor 10^{-11} than the transistors we make today. The energy requirement for a single switch is also about eleven orders of magnitude smaller than the energy required to switch the transistor today, and the time to make the transitions will be at least ten thousand times faster per step of calculation. So there is plenty of room for improvement in the computer and I leave this to you, practical people who work on computers, as an aim to get to. I underestimated how long it would take for Mr. Ezawa to translate what I said, and I have no more to say that I have prepared for today. Thank you! I will answer questions if you'd like.

Questions and Answers

Q: You mentioned that one bit of information can be stored in one atom, and I wonder if you can store the same amount of information in one quark.

A: Yes. But we don't have control of the quarks and that becomes a really impractical way to deal with things. You might think that what I am talking about is impractical, but I don't believe so. When I am talking about atoms, I believe that someday we will be able to handle and control them individually. There would be so much energy involved in the quark interactions that they would be very dangerous to handle because of the radioactivity and so on. But the atomic energies that I am talking about are very familiar to us in chemical energies, electrical energies, and those are numbers that are within the realm of reality, I believe, however absurd it may seem at the moment.

Q: You said that the smaller the computing element is the better. But, I think equipment has to be larger, because. . .

A: You mean that your finger is too big to push the buttons? Is that what you mean?

Q: Yes, it is.

A: Of course, you are right. I am talking about internal computers perhaps for robots or other devices. The input and output is something that I didn't discuss, whether the input comes from looking at pictures, hearing voices, or buttons being pushed. I am discussing how the computation is done in principle and not what form the output should take. It is certainly true that the input and the output cannot be reduced in most cases effectively beyond human dimensions. It is already too difficult to push the buttons on some of the computers with our big fingers. But with elaborate computing problems that take hours and hours, they could be done very rapidly on the very small machines with low energy consumption. That's the kind of machine I was thinking of. Not the simple applications of adding two numbers but elaborate calculations.

Q: I would like to know your method to transform the information from one atomic scale element to another atomic scale element. If you will use a quantum mechanical or natural interaction between the two elements, then such a device will become very close to Nature itself. For example, if we make a computer simulation, a Monte Carlo simulation of a magnet to study critical phenomena, then your atomic scale computer will be very close to the magnet itself. What are your thoughts about that?

A: Yes. All things that we make are Nature. We arrange it in a way to suit our purpose, to make a calculation for a purpose. In a magnet there is some kind of relation, if you wish; there are some kinds of computations going on, just like there are in the solar system, in a way of thinking. But that might not be the calculation we want to make at the moment. What we need to make is a device for which we can change the programs and let it compute the problem that we want to solve, not just its own magnet problem that it likes to solve for itself. I can't use the solar system for a computer unless it just happens that the problem that someone gave me was to find the motion of the planets, in which case all I have to do is to watch. There was an amusing article written as a joke. Far in the future, the "article" appears discussing a new method of making aerodynamical calculations: Instead of using the elaborate computers of the day, the author invents a simple device to blow air past the wing. (He reinvents the wind tunnel!)

Q: I have recently read in a newspaper article that operations of the nerve system in a brain are much slower than present-day computers and the unit in the nerve system is much smaller. Do you think that the computers you have talked about today have something in common with the nerve system in the brain?

A: There is an analogy between the brain and the computer in that there are apparently elements that can switch under the control of others. Nerve impulses controlling or exciting other nerves, in a way that often depends upon whether more than one impulse comes in—something like an AND or its generalization. What is the amount of energy used in the brain cell for one of these transitions? I don't know the number. The time it takes to make a switching in the brain is very much longer than it is in our computers even today, never mind the fancy business of some future atomic computer, but the brain's interconnection system is much more elaborate. Each nerve is connected to thousands of other nerves, whereas we connect transistors only to two or three others.

Some people look at the activity of the brain in action and see that in many respects it surpasses the computer of today, and in many other respects the computer surpasses ourselves. This inspires people to design machines that can do more. What often happens is that an engineer has an idea of how the brain works (in his opinion) and then designs a machine that behaves that way. This new machine may in fact work very well. But, I must warn you that that does not tell us anything about how the brain actually works, nor is it necessary to ever really know that, in order to make a computer very capable. It is not necessary to understand the way birds flap their wings and how the feathers are designed in order to make a flying machine. It is not necessary to understand the lever system in the legs of a cheetah—an animal that runs fast—in order to make an automobile with wheels that goes very fast. It is therefore not necessary to imitate the behavior of Nature in detail in order to engineer a device which can in many respects surpass Nature's abilities. It is an interesting subject and I like to talk about it.

Your brain is very weak compared to a computer. I will give you a series of numbers, one, three, seven... Or rather, ichi, san, shichi, san, ni, go, ni, go, ichi, hachi, ichi, ni, ku, san, go. Now I want you to repeat them back to me. A computer can take tens of thousands of numbers and give me them back in reverse, or sum them or do lots of things that we cannot do. On the other hand, if I look at a face, in a glance I can tell you who it is if I know that person, or that I don't know that person. We do not yet know how to make a computer system so that if we give it a pattern of a face it can tell us such information, even if it has seen many faces and you have tried to teach it.

Another interesting example is chess playing machines. It is quite a surprise that we can make machines that play chess better than almost everybody in the room. But they do it by trying many, many possibilities. If he moves here, then I could move here, and he can move there, and so forth. They look at each alternative and choose the best. Computers look at millions of alternatives, but a master chess player, a human, does it differently. He recognizes patterns. He looks at only thirty or forty positions before deciding what move to make. Therefore, although the rules are simpler in Go, machines that play Go are not very good, because in each position there are too many possibilities to move and there are too many things to check and the machines cannot look deeply. Therefore the problem of recognizing patterns and what to do under these circumstances is the thing that the computer engineers (they like to call themselves computer scientists) still find very difficult. It is certainly one of the important things for future computers, perhaps more important than the things I spoke about. Make a machine to play Go effectively!

Q: I think that any method of computation would not be fruitful unless it would give a kind of provision on how to compose such devices or programs. I thought the Fredkin paper on conservative logic was very intriguing, but once I came to think of making a simple program using such devices I came to a halt because thinking out such a program is far more complex than the program itself. I think we could easily get into a kind of infinite regression because the process of making out a certain program would be more complex than the program itself and in trying to automate the process, the automating program would be much more complex and so on, especially in this case where the program is hard-wired rather than being separated as a software. I think it is fundamental to think of the ways of composition.

A: We have some different experiences. There is no infinite regression: It stops at a certain level of complexity. The machine that Fredkin ultimately is talking about and the one that I was talking about in the quantum mechanical case are both universal computers in the sense that they can be programmed to do various jobs. This is not a hard-wired program. They are no more hard-wired than an ordinary computer that you can put information in—the program is a part of the input—and the machine does the problem that it is assigned to do. It is hard-wired but it is universal like an ordinary computer. These things are very uncertain but I found an algorithm. If you have a program written for an irreversible machine, the ordinary program, then I can convert it to a reversible machine program by a direct translation scheme, which is very inefficient and uses many more steps. Then, in real situations, the number of steps can be much less. But at least I know that I can take a program with $2n$ steps where it is irreversible, convert it to $3n$ steps of a reversible machine. That is many more steps. I did it very inefficiently since I did not try to find the minimum—just one way of doing it. I don't really think that we'll find this regression that you speak of, but you might be right. I am uncertain.

Q: Won't we be sacrificing many of the merits we were expecting of such devices, because those reversible machines run so slow? I am very pessimistic about this point.

A: They run slower, but they are very much smaller. I don't make it reversible unless I need to. There is no point in making the machine reversible unless you are trying very hard to decrease the energy enormously, rather ridiculously, because with only 80 times kT the irreversible machine functions perfectly. That 80 is much less than the present-day 10^9 or 10^{10} kT, so I have at least 10^7 improvement in energy to make, and can still do it with irreversible machines! That's true. That's the right way to go, for the present. I entertain myself intellectually for fun, to ask how far could we go in principle, not in practice, and then I discover that I can go to a fraction of a kT of energy and make the machines microscopic, atomically microscopic. But to do so, I must use the reversible physical laws. Irreversibility comes because the heat is spread over a large number of atoms and can't be gathered back again. When I make the machine very small, unless I allow a cooling element which is lots of atoms, I have to work reversibly. In practice there probably will never come a time when we will be unwilling to tie a little computer to a big piece of lead which contains 10^{10} atoms (which is still very small indeed), making it effectively irreversible. Therefore I agree with you that in practice, for a very long time and perhaps forever, we will use irreversible gates. On the other hand, it is a part of the adventure of science to try to find a limitation in all directions and to stretch the human imagination as far as possible everywhere. Although at every stage it has looked as if such an activity was absurd and useless, it often turns out at least not to be useless.

Q: Are there any limitations from the uncertainty principle? Are there any fundamental limitations on the energy and the clock time in your reversible machine scheme?

A: That was my exact point. There is no further limitation due to quantum mechanics. One must distinguish carefully between the energy lost or consumed irreversibly, the heat generated in the operation of the machine, and the energy content of the moving parts which might be extracted again. There is a relationship between the time and the energy which might be extracted again. But that energy which can be extracted again is not of any importance or concern. It would be like asking whether we should add the mc^2, the rest energy, of all the atoms which are in the device. I only speak of the energy lost times the time, and then there is no limitation. However, it is true that if you want to make a calculation at a certain extremely high speed, you have to supply to the machine parts which move fast and have energy, but that energy is not necessarily lost at each step of the calculation; it coasts through by inertia.

A (to no Q): Could I just say with regard to the question of useless ideas, I'd like to add one more. I waited, if you would ask me, but you didn't. So I will answer it anyway. How would we make a machine of such small dimensions that we have to put the atoms in special places? Today we have no machinery with moving parts whose dimension is extremely small, at the scale of atoms or hundreds of atoms even, but there is no physical limitation in that direction either. There is no reason why, when we lay down the silicon even today, the pieces cannot be made into little islands so that they are movable. We could also arrange small jets so we could squirt the different chemicals on certain locations. We can make machinery which is extremely small. Such machinery will be easy to control by the same kind of computer circuits that we make. Ultimately, for fun again and intellectual pleasure, we could imagine machines as tiny as a few microns across, with wheels and cables all interconnected by wires, silicon connections, so that the thing as a whole, a very large device, moves not like the awkward motions of our present stiff machines but in the smooth way of the neck of a swan, which after all is a lot of little machines, the cells all interconnected and all controlled in a smooth way. Why can't we do that ourselves?

3

Los Alamos from Below

✦

And now a little something on the lighter side—gems about wisecracker (not to mention safecracker) Feynman getting in and out of trouble at Los Alamos: getting his own private room by seeming to break the no-women-in-the-men's-dormitory rule; outwitting the camp's censors; rubbing shoulders with great men like Robert Oppenheimer, Niels Bohr, and Hans Bethe; and the awesome distinction of being the only man to stare straight at the first atomic blast without protective goggles, an experience that changed Feynman forever.

Professor Hirschfelder's flattering introduction is quite inappropriate for my talk, which is "Los Alamos from Below." What I mean from below is although in my field at the present time I'm a slightly famous man, at the time I was not anybody famous at all. I did not even have a degree when I started to work on my stuff associated with the Manhattan Project.* Many of the other people who tell you about Los Alamos

---

*The name given to the gargantuan project to build the first atomic bomb, which began in 1942 and culminated with the bombing of Hiroshima and Nagasaki on August 6 and 9, respectively, 1945. The project was spread over the United States, with units at, for example, the University of Chicago; Hanford, Washington; Oak Ridge, Tennessee; and Los Alamos, New Mexico, where the bombs were built, and which was essentially the headquarters of the whole project. *Ed.*

knew somebody up in some higher echelon of governmental organization or something, people who were worried about some big decision. I worried about no big decisions. I was always flittering about underneath somewhere. I wasn't the *absolute* bottom. As it turns out I did sort of get up a few steps, but I wasn't one of the higher people. So I want you to put yourself in a different kind of condition than the introduction said and just imagine this young graduate student who hasn't got his degree yet, who is working on his thesis. I'll start by saying how I got into the project, and then what happened to me. That's all, just what happened to me during the project.

I was working in my office* one day, when Bob Wilson† came in. I was working—[laughter] what the hell, I've got lots funnier yet; what are you laughing at?—Bob Wilson came in and said that he had been funded to do a job that was a secret and he wasn't supposed to tell anybody, but he was going to tell me because he knew that as soon as I knew what he was going to do, I'd see that I had to go along with it. So he told me about the problem of separating different isotopes of uranium. He had to ultimately make a bomb, a process for separating the isotopes of uranium, which was different from the one which was ultimately used, and he wanted to try to develop it. He told me about it and he said there's a meeting. . . I said I didn't want to do it. He said all right, there's a meeting at three o'clock, I'll see you there. I said it's all right you told me the secret because I'm not going to tell anybody, but I'm not going to do it. So I went back to work on my thesis, for about three minutes. Then I began to pace the floor and think about this thing. The Germans had Hitler and the

---

*at Princeton University.

†Robert Wilson, winner (with Arno Penzias) of the 1978 Physics Nobel Prize for discovering the cosmic microwave background radiation. *Ed.*

possibility of developing an atomic bomb was obvious, and the possibility that they would develop it before we did was very much of a fright. So I decided to go to the meeting at three o'clock. By four o'clock I already had a desk in a room and was trying to calculate whether this particular method was limited by the total amount of current that you can get in an ion beam, and so on. I won't go into the details. But I had a desk, and I had paper, and I'm working hard as I could and as fast as I can. The fellows who were building the apparatus planned to do the experiment right there. And it was like those moving pictures where you see a piece of equipment go bruuuup, bruuuup, bruuuup. Every time I'd look up the thing was getting bigger. And what was happening, of course, was that all the boys had decided to work on this and to stop their research in science. All the science stopped during the war except the little bit that was done in Los Alamos. It was not much science; it was a lot of engineering. And they were robbing their equipment from their research, and all the equipment from different research was being put together to make the new apparatus to do the experiment, to try to separate the isotopes of uranium. I stopped my work also for the same reason. It is true that I did take a six-week vacation after a while from that job and finished writing my thesis. So I did get my degree just before I got to Los Alamos, so I wasn't quite as far down as I led you to believe.

One of the first experiences that was very interesting to me in this project at Princeton was to meet great men. I had never met very many great men before. But there was an evaluation committee that had to decide which way we were going and to try to help us along, and to help us ultimately decide which way we were going to separate the uranium. This evaluation committee had men like Tolman and Smyth and Urey and Rabi and Oppenheimer and so forth on it. And there was

Compton, for example. One of the things I saw was a terrible shock. I would sit there because I understood the theory of the process of what we were doing, and so they'd ask me questions and then we'd discuss it. Then one man would make a point and then Compton, for example, would explain a different point of view, and he would be perfectly right, and it was the right idea, and he said it should be *this* way. Another guy would say well, maybe, there's this possibility we have to consider against it. There's another possibility we have to consider. I'm jumping! He should, Compton, he should say it again, he should say it again! So everybody is disagreeing, it went all the way around the table. So finally at the end Tolman, who's the chairman, says, well, having heard all these arguments, I guess it's true that Compton's argument is the best of all and now we have to go ahead. And it was such a shock to me to see that a committee of men could present a whole lot of ideas, each one thinking of a new facet, and remembering what the other fellow said, having paid attention, and so that at the end the decision is made as to which idea was the best, summing it all together, without having to say it three times, you see? So that was a shock, and these were very great men indeed.

This project was ultimately decided not to be the way that they were going to separate uranium. We were told then that we were going to stop and that there would be starting in Los Alamos, New Mexico, the project that would actually make the bomb and that we would all go out there to make it. There would be experiments that we would have to do, and theoretical work to do. I was in the theoretical work; all the rest of the fellows were in experimental work. The question then was what to do, because we had this hiatus of time since we'd just been told to turn off and Los Alamos wasn't ready yet. Bob Wilson tried to make use of his time by sending me to Chicago to find out all that I could about the bomb and the problems so that we could start to build in our laboratories equipment, counters of various kinds, and so on that would be useful when we got to Los Alamos. So no time was wasted. I was sent to Chicago with the instructions to go to each group, tell them I was going to work with them, have them tell me about a problem to the extent that I knew enough detail so that I could actually sit down and start to work on the problem, and as soon as I got that far go to another guy and ask for a problem, and that way I would understand the details of everything. It was a very good idea, although my conscience bothered me a little bit. But it turned out accidentally (I was very lucky) that as one of the guys explained a problem I said why don't you do it that way and in a half an hour he had it solved, and they'd been working on it for three months. So, I did something! When I came back from Chicago I described the situation—how much energy was released, what the bomb was going to be like and so forth to these fellows. I remember a friend of mine who worked with me, Paul Olum, a mathematician, came up to me afterwards and said, "When they make a moving picture about this, they'll have the guy coming back from Chicago telling the Princeton men all about the bomb, and he'll be dressed in a suit and carry a briefcase and so on—and you're in dirty shirtsleeves and just telling us all about it." But it's a very serious thing anyway and so he appreciated the difference between the real world and that in the movies.

Well, there still seemed to be a delay and Wilson went to Los Alamos to find out what was holding things up and how they were progressing. When he got there he found that the construction company was working very hard and had finished the theater, and a few other buildings because they understood how, but they hadn't gotten instructions clear on how to build a laboratory—how many pipes for gas, how much for water—so he simply stood around and decided how much water, how much gas and so on, and told them to start building the laboratories. And he came back to us—we were all ready to go, you see—and Oppenheimer was having some difficulties in discussing some problems with Groves and we were getting impatient. As far as I understand it from the position I was in, Wilson then called Manley in Chicago and they all got together and decided we'd go out there anyway, even if it wasn't ready. So we all went out to Los Alamos before it was ready. We were recruited, by the way, by Oppenheimer and other people and he was very patient with everybody; he paid attention to everybody's problems. He worried about my wife, who had TB, and whether there would be a hospital out there and everything, and it was the first time I met him in such a personal way and he was such a wonderful man. We were told among other things, for example, to be careful. Not to buy our train ticket in Princeton. Because Princeton was a very small train station, and if everybody bought train tickets to Albuquerque, New Mexico, there would be suspicion that something was up. And so everybody bought their tickets somewhere else, except me, because I figured if everybody bought their tickets somewhere else. . . . So when I went to the train station and I said I want to go to Albuquerque, New Mexico, he says, oh, he says, so all this stuff is for *you*! We had been shipping out crates full of counters for weeks and expecting they didn't notice that the address was Albuquerque. So at least I explained why it was that we were shipping out crates—I was going out to Albuquerque.

Well, when we arrived we were ahead of time and the houses for the dormitories and things like that were not ready. In fact, the laboratories weren't quite ready. We were pushing them, we were driving them by coming down ahead of time. They went crazy at the other end and they rented ranch houses all around in the neighborhood. And we stayed at first in a ranch house and would drive in in the morning. The first morning I drove in was tremendously impressive; the beauty of the scenery, for a person from the East who didn't travel much, was sensational. There are the great cliffs; you've probably seen the pictures, I won't go into it in much detail. These things were high on a mesa and you'd come up from below and see these great cliffs and we were very surprised. The most impressive thing to me was that as I was going up, I said that maybe there were Indians even living here, and the guy who was driving the car just stopped; he stopped the car and walked around the corner and there were Indian caves that you could inspect. So it was really very exciting, in that respect.

When I got to the site the first time, I saw at the gate—you see there was a technical area that was supposed to have a fence around it ultimately, but because they were still building, it was still open. Then there was supposed to be a town and then a *big* fence further out, around the town—my friend Paul Olum, who was my assistant, standing with a clipboard checking the trucks coming in and out and telling them which way to go to deliver the materials in different places. When I went into the laboratory I would meet men I had heard of by seeing their papers in the *Physical Review* and so on. I had never met them before. This is John Williams, they said. A guy comes standing up from a desk which is covered with blueprints, his sleeves all rolled up, and he's standing there by some windows at one of the buildings ordering trucks and things going in different directions to build the things. In other words, we took over the construction company and finished the job. The physicists, in the beginning the experimental physicists particularly, had nothing to do until their buildings were ready, and apparatus was ready, so they just built the buildings, or assisted in building the buildings. The theoretical physicists, on the other hand, it was decided that they wouldn't live in the ranch houses, but they would live up at the site because they could start working right away. So we started working immediately, and that meant that we would each get a roll blackboard, you know, on wheels that you'd roll around, and we'd roll it around and Serber would explain to us all the things that they'd thought of in Berkeley about the atomic bomb, and nuclear physics and all these things, and I didn't know very much about it. I had been doing other kinds of things. And so I had to do an awful lot of work. Every day I would study and read, study and read, and it was a very hectic time. I had some luck. All the big shots had by some kind of accident—everybody but Hans Bethe—happened to have left at the same time; like Weisskopf had to go back to fix something at MIT, and Teller was away, just at a certain moment, and what Bethe needed was someone to talk to, to push his ideas against. Well, he came to this little squirt in an office and he starts to argue, to explain his idea. I said, "No, no, you're crazy, it'll go like this." And he said, "Just a moment," and he explained how he's not crazy, that I'm crazy, and we keep on going like this. It turned out that, although I—you see, when I hear about physics I just think about physics and I don't know who I'm talking to and I say the dopiest things like no, no, you're wrong or you're crazy—but it turned out that's exactly what he needed. So I got a notch up on account of that and I ended up as a group leader with four guys under me, which is underneath Bethe.

I had a lot of interesting experiences with Bethe. The first day when he came in, we had an adding machine, a Marchant that you work with your hands, and so he said "Let's see, the pressure"—the formula which he'd been working out involves the pressure squared—"the pressure is 48; the square of 48. . . ." I reach for the machine; he says it's about 23 hundred. So I plug it out just to find out. He says, "You want to know exactly? It's 2,304." And so it came out 2,304. So I said, "How do you do that?" He says, "Don't you know how to take squares of numbers near 50? If it's near 50, say 3 below, then it's 3 below 25, like 47 squared is 22. And how much is left over is the square of what's residual. For instance, with your 3 less you get 9—2,209 from 47 squared. Very nice, OK?" So we (he was very good at arithmetic) kept on going and a few moments later we had to take the cube root of $2\frac{1}{2}$. Now, to do cube roots there was a little chart that you take that had some trial numbers that you try on the adding machine that the Marchant Company had given us. So (this takes him a little longer, you see) I opened the drawer, take out the chart, and he says "1.35." So I figured there is some way to take cube roots numbers near $2\frac{1}{2}$, but it turns out no. I said, "How do you do that?" He says, "Well," he says "you see the logarithm of 2.5 is so-and-so; you divide by 3 to get the cube root of so and so. Now, the log of 1.3 is this, the log of 1.4 is . . . I interpolate in between." I couldn't have divided anything by three, much less . . . So he knew all his arithmetic and he was very good at it and that was a challenge to me. I kept practicing. We used to have a little contest. Every time we'd have to calculate anything we'd rush to the answer, he and I, and I would win; after several years I began to be able to do it, you know get in there once, maybe one out of four. Of course you'd notice something funny about a number like if you have to multiply 174 by 140, for example. You notice that 173 by 141 which is like the square root of 3, times the square root of 2, which is the square root of 6, which is 245. But you have to notice the numbers, you see, and each guy would notice a different way—we had lots of fun.

Well, when I was first there, as I said, we didn't have the dormitories, and the theoretical physicists had to stay on the site. The first place they put us was in the old school building, from the boys' school which had been there previously. The first place I lived in was a thing called the mechanics lodge; we were all jammed in there in bunk beds and so on and it turned out it wasn't organized very well and Bob Christie and his wife had to go to the bathroom each morning through our bedroom. So that was very uncomfortable.

The next place we moved to was a thing called the Big House, which had a patio all the way around the outside on the second floor where all the beds had been stuck next to each other, all along the wall. And then downstairs was a big chart that told you what your bed number was and which bathroom you changed your clothes in. So under my name it said "Bathroom C," no bed number! As a result of this I was rather annoyed. At last the dormitory is built. I go down to the dormitory place to get rooms assigned, and they say you can pick your room now. I tried to pick one; you know what I did–I looked to see where the girls' dormitory was and I picked one that you could look out across. Later I discovered a big tree was growing right in front of it. But, anyway, I picked this room. They told me that temporarily there would be two people in a room, but that would only be temporary. Two rooms would share a bathroom. It would be double-decker beds, bunks, in there and I didn't want two people in the room. When I first got there, the first night, nobody else was there. Now my wife was sick with TB in Albuquerque, so I had some boxes of stuff of hers. So I opened a box and I took out a little nightgown and I just sort of threw it carelessly. I opened the top bed, I threw the nightgown careless on the top bed. I took out the slippers; I threw some powder on the floor in the bathroom. I just made it look like somebody else was there. OK?

So if the other bed is occupied, nobody else is going to sleep there. OK? So, what happened? Because it's a men's dormitory. Well, I came home that night and my pajamas are folded nicely and put under the pillow, and the slippers put nicely at the bottom of the bed. The lady's pajama is nicely folded and it's been put under the pillow, the bed is all fixed up and made, and the slippers are put down nicely. The powder is cleaned from the bathroom and *nobody* is sleeping up there. I still have the room to myself. So, next night, same thing. When I wake up I mess up the bed up top, I throw the nightgown, and the powder in the bathroom and so on, and I went on like this for four nights until it settled down. Everybody was settled and there was no more danger that they would put a second person in the room. Each night, everything was set very neatly, everything was all right, even though it was a men's dormitory. So, that's what happened in that situation.

I got involved in politics a little bit because there was a thing called the Town Council. Apparently there were certain things the Army people would decide about how the town was supposed to be run, with assistance from some governing board up there that I never knew anything about. But there was all kinds of excitement like there is in any political thing. In particular, there were factions: the housewives' faction, the mechanics' faction, the technical people's faction, and so on. Well, the bachelors and bachelor girls, the people who lived in the dormitory, felt they had to have a faction because a new rule had been promulgated—no women in the men's dorm, for instance. Well, this is absolutely ridiculous. All grown people of course (ha, ha). What kind of nonsense? So we had to have political action. So we decided and we debated and this stuff; you know how it is. And so I was elected to represent the dormitory people, you see, in the Town Council.

◆

The Pleasure of Finding Things Out

After I was in the Town Council for about a year or so, a year and a half, I was talking to Hans Bethe about something. He was up in the governing council during all this time. And I told him this story, I had done this trick one time with my wife's stuff in the upper bed and he starts to laugh. He says, "Ha, that's how you got on the Town Council." Because it turned out that what happened was this. There was a report, a very serious report. The poor woman was shaking, the woman who cleans the rooms in the dormitory had just opened the door and all of a sudden there is trouble—somebody is sleeping with one of the guys! Shaking, she doesn't know what to do. She reports, the charwoman reports to the chief charwoman, the chief charwoman reports to the lieutenant, the lieutenant reports to the major, it goes all the way up, it goes all the way up through to the generals to the governing board—what are they going to do?—they're going to think about it! So in the meantime, what instructions go down, down, through the captains, down through the majors, through the lieutenants, through the chars' chief, through the charwoman?—"Just put things back the way they are, clean 'em up" and see what happens. OK? Next day, report—same thing, brump, bruuuuump, bruuuump. Meantime, for four days, they worried up there, what they're going to do. So they finally promulgated a rule. "No women in the men's dormitory!" And that caused such a *stink* down there. You see, now they had to have all the politics and they elected somebody to represent them. . . .

Now I would like to tell you about the censorship that we had. They decided to do something utterly illegal, which was to censor mail of people inside the United States, in the Continental United States, which they have no right to do. So it had to be set up very delicately, as a voluntary thing. We would all volunteer not to seal our envelopes that we would send our letters out with. We would accept, it would be all right, that they would open letters coming in to us; that was voluntarily accepted by us. We would leave the outgoing letters open; they would seal them, if they were OK. If they weren't OK in their opinion, in other words they found something that we shouldn't send out, they would send the letter back to us with a note that there was a violation of such and such a paragraph of our "understanding," and so on and so forth. So, very delicately, amongst all these liberal-minded scientific guys agreeing to such a proposition, we finally got the censorship set up. With many rules, such as that we were allowed to comment on the character of the administration if we wanted to, so we could write our senator and tell him we don't like the way things are run, and things like that. So it was all set up and they said that they would notify us if there were any difficulties.

So, the day starts, the first day for censorship. Telephone! Briiing! Me—"What?" "Please come down." I come down. "What's this?" It's a letter from my father. "Well, what is it?" There's lined paper, and there's these lines going out with dots—four dots under, one dot above, two dots under, one dot above, dot under dot. "What's that?" I said, "It's a code." They said, "Yes, it's a code; but what does it say?" I said, "I don't know what it says." They said "Well, what's the key to the code; how do you decipher it?" I said, "Well, I don't know." Then they said, "What's this?" I said, "It's a letter from my wife." "It says TJXYWZ TW1X3. What's this?" I said, "Another code." "What's the key to it?" "I don't know." They said, "You're receiving codes and you don't know the key?" "Precisely," I said, "I have a game. I challenge them to send me a code that I can't decipher, see? So they're making up codes at the other end and they're not going to tell me what the key is and they're sending them in." Now one of the rules of the censorship was that they aren't going to disturb anything that you would ordinarily do, in the mail. So they said, "Well, you're going to have to tell them please to send the key in with the code." I said, "I don't want to see the key!" They said, "Well, all right, we'll take the key out." So we had that arrangement. OK? All right. Next day, I get a letter from my wife which says "It's very difficult writing because I feel that blank is looking over my shoulder." And in the spot there is nicely eradicated a splotch with ink eradicator. So I went down to the bureau and I said, "You're not supposed to touch the incoming mail if you don't like it, you can tell me but you're not supposed to do anything to it. Just look at it but you're not supposed to take anything out." They said, "Don't be ridiculous; do you think that's the way censors work anyway, with ink eradicator? They cut things out with the scissor." I said OK. So I wrote a letter back to my wife and said, "Did you use ink eradicator in your letter?" She writes back, "No, I didn't use ink eradicator in my letter; it must have been the ————," and there's a hole cut out. So I went back to the guy in charge, the major who was supposed to be in charge of all this, and complained. This went on for a few days. I felt I was sort of the representative to get the thing straightened out. He tried to explain to me that these people who were the censors had been taught how to do it, and they didn't understand this new way that we had to be so delicate about. I was trying to be the front, the one with the most experience, I was writing back and forth to my wife every day anyway. So, he said, "What's the matter, don't you think I have good faith, have good will?" I says, "Yes, you have perfectly good will but I don't think you have power." 'Cause you see, this had gone on three or four days. He said "We'll see about that!" He grabs the telephone . . . everything was straightened out. So no more was the letter cut.

However, there were a number of difficulties that arose. For example, one day I got a letter from my wife and a note from the censor that said there was a code enclosed, without the key, and so we removed it. So when I went to see my wife in Albuquerque that day, she says "Well, where's all the stuff?" I said, "What stuff?" She says "Lethage, glycerine, hot dogs, laundry." I said, "Wait a minute, that was a list?" She says, "Yes." "That was a *code*," I said. They thought it was a code—lethage, glycerine, etc. Then one day I am jiggling around—in the first few weeks all this went on, it was a few weeks before we got each other straightened out, but—I'm piddling around with the adding machine, the computing machine, and I notice something. So, writing every day, I had a lot of things to write—so, it's very peculiar. Notice what happens. If you take one divided by 273 you get 0.004115226337. It's quite acute and then it goes a little cockeyed when your carrying occurs for only about three numbers and then you can see how the 10 10 13 is really equivalent to 114 again, or 115 again, and it keeps on going, and I was explaining that, how nicely it repeated itself after a couple of cycles. I thought it was kind of amusing. Well, I put that in the mail and it comes back to me; it doesn't go through, and there's a little note "Look at Paragraph 17B." I look at paragraph 17B. It says "Letters are to be written only in English, Russian, Spanish, Portuguese, Latin, German, and so forth. Permission to use any other language must be obtained in writing." And then it said "No codes." So I wrote back to the censor a little note included in my letter which said that I feel that of course this cannot be a code, because if you actually *do* divide 273 by 1 you do, in fact, *get* ———— and I wrote all that there, and therefore there's no more information in the number 1-1-1-1-zero, zero, zero than there is in the number 273 which is hardly any information. And so forth. I therefore asked for permission to write my letters in Arabic numerals. I like to use Arabic numerals in my letters. So, I got that through all right.

There was always some kind of difficulty with the letters going back and forth. At one time my wife kept insisting on mentioning the fact that she feels uncomfortable writing with the feeling that the censor is looking over [her shoulder]. As a rule we aren't supposed to mention censorship—*we* aren't, but how can they tell her? So they keep sending me a note, "Your wife mentioned censorship." *Certainly,* my wife mentioned censorship, so finally they sent me a note that said "Please inform your wife not to mention censorship in her letters." So I take my letter and I start, "I have been instructed to inform you not to mention censorship in your letters." Phoom, phoooom, it comes right back! So I write, "I have been instructed to inform her not to mention censorship. How in the heck am I going to do it? Furthermore, *why* do I have to instruct her not to mention censorship? You keeping something from me?" It is very interesting that the censor himself has to tell me to tell my wife not to tell me that she's . . . But they had an answer. They said, yes, that they are worried about mail being intercepted on the way from Albuquerque, and that they would find out that there was censorship if they looked in the mail and would she please act much more normal. So I went down the next time to Albuquerque and I talked to her and I said, "Now, look, let's not mention censorship," but we had had so much trouble that we had at last worked out a code, something illegal. We had a code; if I would put a dot at the end of my signature it meant I had had trouble again, and she would move on to the next of the moves that she had concocted. She would sit there all day long because she was ill and she would think of things to do. The last thing that she did was to send me, which she found perfectly legitimately, an advertisement that said "Send your boyfriend a letter on a jigsaw puzzle. Here are the blanks. We sell you the blank, you write the letter on it, take it all apart, you put it in a little sac and mail it." So I received that one with a note saying "We do not have time to play games. Please instruct your wife to confine herself to ordinary letters!" Well, we were ready with the one more dot. The letter would start, "I hope you remembered to open this letter carefully because I have included the Pepto-Bismol for your stomach as we arranged." It would be a letter full of powder. In the office we expected they would open it quickly, the powder would go all over the floor, they would get all upset because you are not supposed to upset anything, you'd have to gather all this Pepto-Bismol. . . . But we didn't have to use that one. OK?

As a result of all these experiences with the censor, I knew exactly what could get through and what could not get through. Nobody else knew as well as I. And so I made a little money out of all of this by making bets. One day, on the outside fence, I had discovered that workmen who lived still further out and wanted to come in were too lazy to go around through the gate, and so they had cut themselves a hole, some distance along. So I went out the fence, went over to the hole and came in, went out again, and so on, until the guy, the sergeant at the gate begins to wonder what's happening, this guy is always going out and never coming in? And of course his natural reaction was to call the lieutenant and try to put me in jail for doing this. I explained that there was a hole. You see, I was always trying to straighten people out, point out that there was a hole. And so I made a bet with somebody that I could tell where the hole in the fence was, in the mail, and mail it out. And sure enough, I did. And the way I did it was I said—"You should see the way they administer this place"; you see, that's what we were allowed to say. "There's a hole in the fence 71 feet away from such and such a place, that's this size and that size, that you can walk through." Now, what can they do? They can't say to me that there is no such hole. I mean, what are they going to do, it's their own hard luck that there's such a hole. They should *fix* the hole. So, I got that one through. I also got through a letter that told about how one of the boys who worked in one of my groups had been wakened up in the middle of the night and grilled with lights in front of him by some idiots in the Army, because they found out something about his father or something. I don't know, he was supposed to be a communist. His name was Kamane. He's a famous man now.

Well, there was also some other things. I was always trying to straighten them out, like point out the holes in the fence and so forth, but I was always trying to point these things out in a non-direct manner. And one of the things I wanted to point out was this: that at the very beginning we had terribly important secrets. We'd worked out lots of stuff about uranium, how it worked, and all this stuff was in documents that were in filing cabinets that were made out of wood that had on them little ordinary, common padlocks. Various things made by the shop were on the cabinets, like a rod that would go down and then a padlock to hold it, but it was always just a padlock. Furthermore, you could get the stuff, without even opening the padlock, out of these wooden cabinets; you just tilt it over backwards and the bottom drawer, you know, has a little rod that's supposed to hold it. And there's a hole in the wood underneath; you can pull the papers out from below. So I used to pick the locks all the time and point out that it was very easy to do. And every time we had a meeting of the whole group, everybody together, I would get up and I'd say that we have important secrets and we shouldn't keep them in such things. These were such poor locks. We need better locks. And so one day Teller got up at the meeting, and said to me, "Well, I don't keep my most important secrets in my filing cabinet; I keep them in my desk drawer. Isn't that better?" I said, "I don't know, I haven't seen your desk drawer." Well, he's sitting near the front of the meeting and I'm sitting further back. So the meeting continues and I sneak out of the meeting and I go down to see his desk drawer. OK? I don't even have to pick the lock on the desk drawer. It turns out that if you put your hand in the back underneath you can pull out the paper like those toilet paper dispensers; you pull out one, it pulls another, it pulls another. . . . I emptied the whole damn drawer, took everything out, and put it away to one side and then went up on the higher floor and came back. The meeting is just ending and everybody is just coming out and I join the crew like this, you see, walking along with it, and run up to catch up with Teller, and say, "Oh, by the way, let me see your desk drawer." So he says, "Certainly," so we walk into his office and he shows me the desk and I look at it and say that looks pretty good to me. I said "Let's see what you have in there." "I'd be very glad to show it to you," he says, putting in the key and opening the drawer, "if you hadn't already seen it yourself." The trouble with playing a trick on a highly intelligent man like Mr. Teller is the *time* it takes him to figure out from the moment that he sees there is something wrong till he understands exactly what happened is too damn small to give you any pleasure!

Well, I had a lot of other fun with the safes but it has nothing to do with Los Alamos, so I won't discuss it further. I want to tell about some of the problems, special problems, that I had that are rather interesting. One thing had to do with the safety of the plant at Oak Ridge. Los Alamos was going to make the bomb, but at Oak Ridge they were trying to separate the isotopes of uranium, uranium 238 and uranium 236, 235, the latter one, which was the explosive one, all right? So, they were *just* beginning to get infinitesimal amounts from an experimental thing, of 235, and at the same time they were practicing. There was a big plant, they were going to have vats of the stuff, chemicals, and they were going to take the purified stuff and repurify and get it ready for the next stage. You have to purify it in several stages. So they were practicing the chemistry on the one hand and they were just getting a little bit from one of the pieces of apparatus experimentally on the other hand. And they were trying to learn how to assay it, to determine how much uranium 235 there is in it, and we would send them instructions and they never got it right. So finally Segrè* said that the only possible way to get it right was that he'd have to go down there to see what they're doing, to understand why the assay was wrong. The Army people said no, it is our policy to keep all the information of Los Alamos at one place, and that the people in Oak Ridge should not know anything about what it was used for; they just knew what they were trying to do. I mean the higher people knew they were separating uranium, but they didn't know how powerful the bomb was or exactly how it worked or anything. And the people underneath didn't know at *all* what they were doing. And the Army wanted to keep it that way; there was no information going back and forth, but Segrè finally insisted on it, that it was important. They'd never get the assays right, the whole thing would go up in smoke. So Segrè went down to see what they were doing and as he was walking through he saw them wheeling a tank carboy of water, green water; the green water is uranium nitrate. He says, "You're going to handle it like that when it's purified too? Is that what you're going to do?" They said, "Sure, why not?" "Won't it explode?" he says. "Huh?! *Explode!??*" And so the

---

*Emilio Segrè, winner (with Owen Chamberlain) of the 1959 Physics Nobel Prize for discovering the antiproton. *Ed.*

Army said, "You see, we shouldn't have let any information go across!" Well, it turned out that the Army had realized how much stuff we needed to make a bomb, 20 kilograms or whatever it was, and they realized that that much material would never, purified, would never be in the plant, so there was no danger. But they did not know that the neutrons were enormously more effective when they are slowed down in water. And so in water it takes less than a tenth, no a hundredth, very much less material to make a reaction which makes radioactivity. It doesn't make a big explosion, but it makes radioactivity, it kills people all around and so on. So, it was very dangerous and they had not paid any attention to the safety at all.

So a telegram goes from Oppenheimer to Segrè: Go through the entire plant, notice where all the concentrations are supposed to be, with the process as *they* designed it. We will calculate in the meantime how much material can come together before there's an explosion. And so two groups started working on it. Christie's group worked on water solutions and I worked on dry powder in boxes, my group. And we calculated about how much material. And Christie was going to go down and tell them all at Oak Ridge what the situation was. And so I happily gave all my numbers to Christie, and said, you have all the stuff, and go. Christie got pneumonia; I had to go. I never traveled on an airplane before; I traveled on an airplane. They *strapped* the secrets, with a little thing with a belt, on my back! The airplane in those days was like a bus. You stop off every once in a while except the stations were further apart. You stop off to wait. There's a guy standing there next to me with a keychain, swinging it, saying something like, "It must be *terribly* difficult to fly without a priority on airplanes these days." I couldn't resist. I said, "Well, I don't know, I *have* a priority." A little bit later some generals come aboard and they are going to put out some of us number 3s. It's all right, I'm a number 2. That passenger probably wrote to his congressman, if he wasn't a congressman himself, saying, what are they doing sending these little kids around with number priorities in the middle of the war? At any rate, I arrived there. The first thing I did was have them take me to the plant and I said nothing; I just looked at everything. I found out that the situation was even worse than Segrè reported because he was confused at the first time through. He noticed certain boxes in big lots but he didn't notice other boxes in another room in a big lot, but it was the same room on the other side. And things like that. So if you have too much stuff together it goes up, you see. So I went through the entire plant and, I have a very bad memory but when I work intensively, I have a good short-term memory and so I could remember all kinds of crazy things like building ninety-two-oh-seven, vat number so and so, and so forth. I went home that night and I went through the whole thing explaining where all the dangers were, what you would have to do to fix this. It's rather easy—you put cadmium in solutions to absorb the neutrons in the water, you separate the boxes so they are not too dense, too much uranium together and so on, according to certain rules. And so I used all the examples, worked out all the examples and how the process of freezing worked. I felt that you couldn't make the plant safe unless you knew how it worked. So the next day there was going to be a big meeting.

Oh, I forgot to say, before I left, Oppenheimer said to me, "Now," he said, "when you go, the following people are technically able down there at Oak Ridge: Mr. Julian Webb, Mr. So and so, and so on. I want you to make sure that these people are at the meeting, that you tell them how the thing, you know, the safety, that they really *understand*—they're in charge." I said, "What if they're not at the meeting, what am I supposed to do?" He said, "Then you should say—*Los*

Alamos cannot accept the responsibility for the safety of the Oak Ridge plant unless!!!" I said, "You mean me, little Richard's, going to go in there and say...?" He says, "Yes, little Richard, you go and do that." I really grew up fast! So when I arrived, sure enough, I arrived there and the meeting was the next day and all these people from the company, the big shots in the company and the technical people that I wanted were there, and the generals and so forth, that were interested in the problems, organizing everything. It was a big meeting about this very serious problem of safety, because the plant would never work. It would have blown up, I swear it would have, if nobody had paid attention. So there was a lieutenant who took care of me. He told me that the colonel said that I shouldn't tell them how the neutrons work and all the details because we want to keep the things separate. Just tell them what to do to keep it safe. I said, in my opinion, it is impossible for them to understand or to obey a bunch of rules if they don't understand, unless they understand how it works. So it's my opinion that it's only going to work if I tell them, and *Los Alamos cannot accept the responsibility for the safety of the Oak Ridge plant unless they are fully informed as to how it works!!* It was great. So he goes to the colonel. "Give me just five minutes," the colonel says. He goes to the window and he stops and thinks and that's what they're very good at. They are good at making decisions. I thought it was very remarkable how a problem of whether or not information as to how the bomb works should be in the Oak Ridge plant had to be decided and could be decided in five minutes. So I have a great deal of respect for these military guys because I never can decide anything very important in any length of time, at all.

So, in five minutes, he says, all right, Mr. Feynman, go ahead. So I sat down and I told them all about neutrons, how they worked, da da, ta ta ta, there are too many neutrons together, you got to keep the material apart, cadmium absorbs, and slow neutrons are more effective than fast neutrons, and yak yak—all stuff which was elementary primer stuff at Los Alamos, but they had never heard of any of it, so I turned out to be a tremendous genius to them. I was a god coming down from the sky! There was all these phenomena that were not understood and never heard of before, I knew all about it, I could give them facts and numbers and everything else. So, from being rather primitive back there at Los Alamos, I was a super genius at the other end. Well, the result was that they decided, they made little groups, to make their own calculations to learn how to do it. They started to redesign plants. The designers of the plants were there, the construction designers, engineers, chemical engineers for the new plant that was going to handle the separated material were there. And other people were there. And I went away again. They told me to come back in a few months; they were going to redesign their plant for the separation.

So I came back in a few months, a month or so, and Stone and Webster Company, the engineers, had finished the design of the plant and now it was for me to look at the plant. OK? How do you look at a plant that ain't built yet? I don't know. So I go into this room with these fellows. There was always a Lieutenant Zumwalt that was always coming around with me, taking care of me, you know; I had to have an escort everywhere. So he goes with me, he takes me into this room and there are these two engineers and a *loooooooong* table, great big long table, tremendous table, covered with a blueprint that's as big as a table; not one blueprint, but a stack of blueprints. I took mechanical drawing when I was in school, but I wasn't too good at reading blueprints. So they start to explain it to me because they thought I was a genius. And they start out, "Mr. Feynman, we would like you to understand, the plant is so designed, you see one of the things we had to avoid was accumulation." Problems like—there's an evaporator working which is trying to accumulate the stuff; if the valve gets stuck or something like that and they accumulate too much stuff, it'll explode. So they explained to me that this plant is designed so that no *one* valve, if any one valve gets stuck nothing will happen. It needs at least two valves everywhere. So then they explain how it works. The carbon tetrachloride comes in here, the uranium nitrate from here comes in here, it goes up and down, it goes up through the floor, comes up through the pipes, coming up from the second floor, bluuuuurp, from the blueprints, down, up, down, up, very fast talking explaining the very, very complicated chemical plant. I'm completely dazed, worse, I don't know what the symbols on the blueprint mean! There is some kind of a thing that at first I think it's a window. It's a square with a little cross in the middle, all over the damn place. Lines with this damn square, lines with the damn square. I think it's a window; no, it can't be a window, 'cause it ain't always at the edge. I want to ask them what it is. You must have been in a situation like this—you didn't ask them right away, right away it would have been OK. But they've been talking a little bit too long. You hesitated too long. If you ask them now they'll say, what are you wasting my time all this time for? I don't know what to do; I think to myself, often in my life I have been lucky. You are not going to believe this story, but I swear it's absolutely true; it's such sensational luck. I thought what am I going to do, what am I going to *do*????? I got an idea. Maybe it's a valve? So, in order to find out whether it's a valve or not I take my finger and I put it down in the middle of one of the blueprints on page number 3 down in the end and I said, "What happens if this valve gets stuck?" figuring they're going to say, "That's not a valve, sir, that's a window." So one looks at the other and says, "Well, if that valve gets stuck," and they go up and down on the blueprint, up and down, the other guy up and down, back and forth, back and forth, and they both look at each other and they turn around to me and they open their mouths— "You're absolutely right, sir." So they roll up the blueprints and away they went and we walked out. And Lt. Zumwalt, who had been following me all the way through, said, "You're a genius. I got the idea you were a genius when you went through the plant once and you could tell them about evaporator C-21 in building 90-207 the next morning," he says, "but what you have just done is so *fantastic*, I want to know how, *how* do you do something like that?" I told him, you try to find out whether it's a valve or not.

Well, another kind of problem that I worked on was this. We had to do lots of calculations and we did them on Marchant calculating machines. By the way, just to give you an idea of what Los Alamos was like, we had these Marchant computers. I don't know if you know what they look like, hand calculators with numbers and you push them, and they multiply, divide, add, and so on. Not like they do easy now, but hard; they were mechanical gadgets. And they had to be sent back to the factory to be repaired. We didn't have a special man to do it, which was the standard way, and so they would always be sent to the factory. Pretty soon you were running out of machines. So I and a few other fellows started to take the covers off. We weren't supposed to—the rules "You take the covers off, we cannot be responsible . . ." So we took the covers off and we had a nice series of lessons. Like the first one we took the cover off, there was a shaft with a hole in it and a spring which was hanging this way, and obviously the spring went in the hole—so that was easy. So anyway, we got like a series of lessons, by God, on how to fix them and we got better and better and we made more and more elaborate repairs. When we got something too complicated we sent it out, back to the factory, but we'd do the easy ones and kept the things going. I also did some typewriters. I ended up doing all the computers; the other fellows quit on me. I did a few typewriters. There was a guy in the machine shop who was better than I was and he took care of typewriters; I took care of adding machines. However, we decided that the big problem was to figure out exactly what happened during the bomb's explosion when you push the stuff in by an explosion and then it goes out again. Exactly what happens, so you can figure out exactly how much energy was released and so on, required much more calculating than we were capable of. And a rather clever fellow by the name of Stanley Frankle realized that it could possibly be done on IBM machines. The IBM company had machines for business purposes, adding machines that are called tabulators for listing sums and a multiplier, just a machine, a big box, you put cards in and it would take two numbers from a card and multiply it and print it on a card. And then there were collators and sorters and so on. So he decided, he'd figured out a nice program. If we got enough of these machines in a room, we would take the cards and put them through a cycle; everybody who does numerical calculations now knows exactly what I'm talking about, but this was kind of a new thing: mass production with machines.

We had done things like this on adding machines. Usually you go one step across yourself, doing everything. But this was different—where you go first to the adder, then we go to the multiplier, then you go to the adder, and so on. So, he designed this thing and ordered the machine from the IBM company, 'cause we realized it was a good way of solving our problems. We found that there was somebody in the Army that had

IBM training. We needed a man to repair them, to keep them going and everything. And they were going to send this fellow, but it was delayed, always delayed. Now, we *always* were in a hurry. I have to explain that—*everything* we did, we tried to do as quickly as possible. In this particular case, we worked out all the numerical steps that we were supposed to, that the machines were supposed to do, multiply this, and then do this, and subtract that. And then we worked out the program, but we didn't have any machine to test it on. So what we did, I arranged, was a room with girls in it, each one had a Marchant. But *she* was the multiplier and *she* was the adder, and this one cubed, and so we had cards, index cards, and all she did was cube this number and send it to the next one. She was imitating the multiplier, the next one was imitating the adder; we went through our cycle, we got all the bugs out. Well, we did it that way. And it turned out that the speed at which we were able to do it—we'd never done mass production calculating and everybody who'd ever calculated before, every single person, did all the steps. But Ford had a good idea, the damn thing works a hell of a lot faster the other way and we got speed with this system that was the predicted speed for the IBM machine, the same. The only difference is that the IBM machines didn't get tired and could work three shifts. But the girls got tired after a while. So, anyway, we got the bugs out during that process and finally the machines arrived, but not the repairman. So, we went down to put them together. And they were one of the most complicated machines of the technology of those days, these computing machines, big things that came partially disassembled with lots of wires and blueprints of what to do. We went down, we put them together, Stan Frankle and I and another fellow, and we had our troubles. Most of the trouble was the big shots coming all the time and saying you're going to break something, going to break something. We put them together and sometimes they would work, and sometimes they were put together wrong and they didn't work. And so we fiddled around and got them to work. We didn't get them all to work and I was last working on some multiplier, saw a bent part inside and I was afraid to straighten it because it might snap off. They were always telling us we were going to bust it irreversibly. And finally the man from the IBM company came, as a matter of fact, according to schedule, but he came and he fixed the rest that we hadn't got ready, and we got the program going. But he had trouble with the one that I had had trouble with. So after three days he was still working on that last one. I went down, I said, "Oh, I noticed that was bent." He said, "Oh, of course, that's all there is to it!" (Snap)—It was all right. So that was it.

Well, Mr. Frankle started this program and began to suffer from a disease, the computer disease, that anybody who works with computers now knows about. It's a very serious disease and it interferes completely with the work. It was a serious problem that we were trying to do. The disease with computers is you *play* with them. They are so wonderful. You have these x switches that determine, if it's an even number you do this, if it's an odd number you do that, and pretty soon you can do more and more elaborate things if you are clever enough, on one machine. And so after a while it turned out the whole system broke down. He wasn't paying any attention; he wasn't supervising anybody. The system was going very, very slowly. The real problem was that he was sitting in a room figuring out how to make one tabulator automatically print arc-tangent x, and then it would start and it would print columns and then bitsi, bitsi, bitsi and calculate the arc-tangents automatically by integrating as it went along and make a whole table in one operation. Absolutely useless. We *had* tables of arc-tangents. But if you've ever worked with computers you understand the disease. The *delight* to be able to see how much you can do. But he got the disease for the first time, the poor fellow who invented the thing got the disease.

And so, I was asked to stop working on the stuff I was doing in my group and go down and take over the IBM group. I noticed the disease and I tried to avoid it. And although they did three problems in nine months, I had a very good group. The first problem was that they had never told the fellows—they had selected from all over the country, a thing called Special Engineer Detachment. There were clever boys from high school, who had engineering ability, and the Army collected them together in the Special Engineer Detachment. They sent them up to Los Alamos. They put them in barracks and they would tell them *nothing*. Then they came to work and what they had to do was work on IBM machines, punching holes, numbers that they didn't understand, nobody told them what it was. The thing was going very slowly. I said that the first thing there has to be is that the technical guys know what we're doing. Oppenheimer went and talked to the security people and got special permission. So I had a nice lecture in which I told them what we were doing, and they were all excited. We're fighting a war. We see what it is. They knew what the numbers meant. If the pressure came out higher, that meant there was more energy released and so on and so on. They knew what they were doing. *Complete* transformation! *They* began to invent ways of doing it better. They improved the scheme. They worked at night. They didn't need supervising in the night. They didn't need anything. They understood everything. They invented several of the programs that we used and so forth. So my boys really came through and all that had to be done was to tell them what it was, that's all. It's just, don't tell them, they're punching holes, please. As a result, although it took them nine months to do three problems before, we did nine problems in *three* months, which is about nearly ten times as fast. But one of the secret ways that we did our problems was this: The problems consisted of a bunch of cards which had to go through a cycle. First add, then multiply, and so it went through the cycle of machines in this room, slowly about as it went around and around. So we figured a way, by taking a different colored set of cards, to put them through a cycle too, but out of phase. We'd do two or three problems at a time. You see this was another problem. While this one was adding, it was multiplying on the other problem. And such managerial schemes, we got many more problems.

Finally, near the end of the war, just before we had to make a test in Alamogordo, the question was, how much energy would be released? We had been calculating the release from various designs but the specific design which was ultimately used we hadn't computed. So Bob Christie came down and said, we would like the results for how this thing is going to work in one month, or some very short time, I don't know, less than that, three weeks. I said, "It's impossible." He said, "Look, you're putting out so and so many problems a week. It takes only two weeks per problem, or three weeks per problem." I said, "I know, it takes much longer to do the problem, but we're doing them in *parallel*. As they go through it takes a long time and there's no way to make it go around faster." So he went out. I began to think—is there a way to make it go around faster? Well, if we did nothing else on the machine, so there was nothing else interfering, and so on and so on. I began to think. I put on the blackboard a challenge—CAN WE DO IT? to the boys. They all respond, yes, we'll work double shifts, we'll work overtime, all this kind of thing, we'll *try* it. We'll *try* it!! And so the rule was, all other problems *out*. Only one problem and just concentrate on this thing. So they started to work.

My wife died in Albuquerque and I had to go down. I borrowed Fuchs' car; he was a friend of mine in the dormitory. He had an automobile. He was using the automobile to take the secrets away, you know, they went down to Santa Fe. He was the spy; I didn't know that. I borrowed his car to go to Albuquerque. The damned thing got three flat tires on the way. I came back from there and I went into the room, because I was supposed to be supervising everything, but I couldn't do it for three days. It was in this *mess,* this big rush to get the answer for the test that was going to be done in the desert. I go into the room and there are three different color cards. There's white cards, there's blue cards, there's yellow cards and I start to say, "Well, you're not supposed to do more than one problem—only one problem!" They said, "Get out, get out, get out. Wait, we'll explain everything." So I waited, and what happened was this. As they went through, sometimes the machine made a mistake or they put a wrong number in; that happened. What we used to have to do was to go back and do that over again. But they noticed this, that the deck represented positions and depth in the machine, in space or something. A mistake made here, in one cycle, only affects the nearby numbers, the next cycle affects the nearby numbers, and so on. It works its way through the pack of cards. If you have fifty cards and you make a mistake at card number 39, it affects 37, 38, and 39. The next card 36, 37, 38, 39, and 40. The next time it spreads like a disease. So they found an error, back a way, and they got an idea. They would only compute a small deck, of ten cards, around the error. And because ten cards could be put through the machine faster than the deck of fifty cards, they would go with this other deck rapidly through while they continued with the fifty cards with the disease spreading. But the other thing was computing faster and they would seal it all up and correct it.

OK? Very clever. That was the way those guys worked, really hard, very clever, to get speed. There was no other way. If they had to stop to try to fix it, we'd have lost out time. We couldn't have got it. That was what they were doing. Of course you know what happened while they were doing that. They found an error in the blue deck. And so they had a yellow deck with a little fewer cards; it was going around faster than the blue deck, you know. Just when they are going crazy, because after they get it straightened out they got to fix the white one, they got to take the other cards out and replace it by the right ones, and continue correctly, and it's rather confusing—you know how those things always are. You don't want to make a mistake. Just at the time when they've got these three decks going, they're trying to seal everything up, the BOSS comes walking in. "Leave us alone," they said, so I left them alone and everything came out; we solved the problem in time and that's the way it was.

I would like to tell you just a few words about some of the people that I met. I was an underling at the beginning. I became a group leader, but I met some very great men—besides the men on the evaluation committee, the men that I met in Los Alamos. And there are so many of them that it's one of my great experiences in life to have met all these wonderful physicists. Men that I had heard of, smaller and larger, but the greatest ones were there also. There was of course Fermi.* He came down once. The first time that he came was from Chicago to consult a little bit, to help us if we had some problems. We had a meeting with him and I had been doing some calculations and gotten some results. The calculations were so elaborate it

---

*(1901–1954) Winner of the 1938 Nobel Prize in Physics for demonstrating the existence of new radioactive substances produced by neutron irradiation and related work. Fermi was also responsible for the first controlled nuclear reaction at the University of Chicago in December 1942. *Ed.*

was very difficult. Now, usually, I was the expert at this, I could always tell you what the answer was going to look like or when I got it I could explain why. But this thing was so complicated I couldn't explain *why* it was like that. So I said to Fermi that I was doing this problem and I started to calculate—he said, wait, before you tell me the result, let me think. It's going to come out like this (he was right), and it's going to come out like this because of so and so. And there's a perfectly obvious explanation. . . . So *he* was doing what I was supposed to be good at, ten times better. So that was quite a lesson to me.

Then there was Von Neumann, who was the great mathematician. He suggested, I won't go into the things here, some very clever technical observations. We had some very interesting phenomena in the computing of the numbers. The problem looked as if it was unstable and he explained why and so forth. It was very good technical advice. But we used to go for walks often to get rest, like on Sunday or something. We'd walk in the canyons in the neighborhood and we'd often walk with Bethe, Von Neumann, and Bacher. It was a great pleasure. And the one thing that Von Neumann gave me was an idea that he had which was interesting. That you don't have to be responsible for the world that you're in, and so I have developed a very powerful sense of social irresponsibility as a result of Von Neumann's advice. It's made me a very happy man since. But it was Von Neumann who put the seed in which grew now into my *active* irresponsibility!

I also met Niels Bohr.* That was interesting. He came down, his name was Nicholas Baker in those days and he came with Jim Baker, his son, whose name is really Aage†. They came

---

*(1885–1962) Winner of the 1922 Nobel Prize in Physics for his work on the structure of atoms and of the radiation emanating from them. *Ed.*

†Aage Bohr (1922–) winner (with Ben Mattelson and James Rainwater)

from Denmark and they came to visit, and they were *very* famous physicists, as you all know. All the big shot guys, to them, he was even a great god; they were listening to him and so on. And he would talk about things. We were at a meeting and everybody wanted to *see* the great Bohr. So there were a lot of people and I was back in a corner somewhere and we talked about, discussed, the problems of the bomb. That was the first time. He came and he went away and all I could see of him was from between somebody's heads, from the corner. Next time he's due to come, in the morning of the day he's due I get a telephone call. "Hello, Feynman?" "Yes." "This is Jim Baker"; it's his son. "My father and I would like to speak to you." "Me? I'm Feynman, I'm just a . . ." "That's right. OK." So, eight o'clock in the morning, before anybody's awake, I go down to the place. We go into an office in the technical area and he says, "We have been thinking how we could make the bomb more efficient and we think of the following idea." I say, "No, it's not going to work, it's not efficient, blah, blah, blah." So he says, "How about so and so?" I said, "That sounds a little bit better, but it's got this damn fool idea in it." So forth, back and forth. I was always *dumb* about one thing, I never knew who I was talking to. I was always worried about the physics; if the idea looked lousy, I said it looked lousy. If it looked good, I said it looked good. Simple proposition, I've always lived that way. It's nice, it's pleasant, if you can do it. I'm lucky. Just as lucky as I am with that blueprint, I'm lucky in my life that I can do that. So this went on for about two hours of going back and forth over lots of ideas, and then tearing back and forth, arguing. The great Niels always lighting his pipe; perpetually, it always went out. And he

---

of the 1975 Nobel Prize in Physics for their theory of the structure of the atomic nucleus. *Ed.*

talked in a way that was un-understandable. He said, "Mumble, mumble," hard to understand, but his son I could understand better. Finally he said, "Well," he says, lighting his pipe, "I guess we can call in the big shots *now*." So then they called all the other guys and had a discussion with them. And then the son told me what had happened was—the last time he was there he said to his son—"Remember the name of that little fellow in the back over there? He's the only guy who's not afraid of me, and will say when I've got a crazy idea. So *next* time when we want to discuss ideas, we're not going to be able to do it with these guys who say everything is yes, yes, Dr. Bohr. Get that guy first, we'll talk with him first."

The next thing that happened was, of course, the test after we'd made the calculations. We had to make the test. I was actually at home on a short vacation at that time, I guess because my wife died, and so I got a message from Los Alamos that said, "The baby will be born, is expected on such and such a day." So I flew back, and I *just* arrived on the site while the buses were leaving; I couldn't even get into my room. At Alamogordo we waited out there in the distance; we were 20 miles away. And we had a radio and they were supposed to tell us when the thing was going to go off and so forth. The radio wouldn't work, and we never knew what was happening. But just a few minutes before it was supposed to go off, the radio started to work and they told us there was 20 seconds or something to go. For people who were far away like we were—others were closer, six miles away—they gave out dark glasses that you could watch it with. Dark glasses!! Twenty miles away from the damn thing, you get dark glasses—you couldn't see a damn thing through dark glasses. So then I figured the only thing that could really hurt your eyes—bright light can never hurt your eyes—it's ultraviolet light that does. So I got behind a truck windshield, so the ultraviolet can't go through glass, so that would be safe, and so I could *see* the damn thing. Other people were never going to *see* the damn thing. OK. Time comes, and this *trememdous* flash out there, so bright I quickly see this purple splotch on the floor of the truck. I said, "That ain't it. That's an after-image." So I turn back up and I see this white light changing into yellow and then into orange. The clouds form and then they disappear again, the compression and the expansion forms and makes clouds disappear. Then finally, a big ball of orange, the center that was so bright, became a ball of orange that started to rise and billow a little bit and get a little black around the edges and then you see it's a big ball of smoke with flashes on the inside of the fire going out, the heat. I saw all that and all this that I just described in just a moment; took about one minute. It was a series from bright to dark and I had seen it. I am about the only guy that actually looked at the damn thing, the first Trinity Test. Everybody else had dark glasses. The people at six miles couldn't see it because they were all told to lie on the floor with their eyes covered, so no-body saw it. The guys up where I was all had dark glasses. I'm the only guy who saw it with the human eye. Finally, after about a minute and a half, there's suddenly a tremendous noise, *BANG,* and then rumble, like thunder, and that's what convinced me. Nobody had said a word during this whole minute, we were all just watching quietly, but this sound re-leased everybody, released me particularly because the solid-ity of the sound at that distance meant that it had really worked. The man who was standing next to me said, when the sound went off, "What's that?" I said, "That was the bomb." The man was William Laurence of the *New York Times*, who had come. He was going to write an article that was going to describe the whole situation. I had been the one who was supposed to have taken him around. It was found that it was too technnical for him.

Later Mr. Smyth of Princeton came and I showed him around Los Alamos. For example, we went into a room and there on the end of a pedestal, a little narrower than that, was a small ball about so big, silver plated—you could put your hand on it, it was warm. It was radioactive; it was plutonium. And we stood at the door of this room talking about it. There was a new element that was made by man that had never existed on the earth before, except for a very short period possibly, at the very beginning. And here it was all isolated and radioactive and had these properties. And we had made it. And so it was very, tremendously valuable, nothing more valuable and so forth and so on. Meanwhile, you know how people do when you talk, you kind of jiggle around the jiggle and so forth. He's kicking the doorstop, you see, and I says, yes, and I said the doorstop is more appropriate than the door. The doorstop was a hemisphere, yellowish metal, gold as a matter of fact. It was a gold hemisphere about so big. What had happened was we needed to do an experiment to see how many neutrons were reflected by different materials in order to save the neutrons so we didn't use so much plutonium. We had tested many different materials. We had tested platinum, we had tested zinc, we had tested brass, we had tested gold. So in making the tests with the gold, we had these pieces of gold and somebody had the clever idea to use that great ball of gold for a doorstop for the door that contained the plutonium, which is quite appropriate.

After the thing went off and we heard about it, there was tremendous excitement at Los Alamos. Everybody had parties, we all ran around. I sat on the end of a jeep and beat drums and so on. Except for one man that I remember. [It] was Bob Wilson, who got me into it in the first place. He's sitting there moping. I said, "What are you moping about?" He said, "It's a terrible thing that we made." I said, "But you started it, you got us into it." You see, what happened to me, what happened to the rest of us is we *started* for a good reason but then we're working very hard to do something, and to accomplish it, it's a pleasure, it's excitement. And you stop to think, you know, you just stop. After you thought at the beginning, you just stop. So he was the only one who was still thinking about it, at that particular moment. I returned to civilization shortly after that and went to Cornell to teach, and my first impression was a very strange one and I can't understand it anymore but I felt very strongly then. I'd sat in a restaurant in New York, for example, and I looked out at the buildings and how far away, I would think, you know, how much the radius of the Hiroshima bomb damage was and so forth. How far down there was down to 34th Street? All those buildings, all smashed and so on. And I got a very strange feeling. I would go along and I would see people building a bridge. Or they'd be making a new road, and I thought, they're *crazy*, they just don't understand, they don't understand. Why are they making new things, it's so useless? But fortunately, it's been useless for 30 years now isn't it, almost, maybe we'll make 30 years. So I've been wrong for 30 years about its being useless making bridges and I'm glad that those other people were able to go ahead. But my first reaction after I was finished with this thing was it's useless to make anything. Thank you very much.

Question: What about your story about some safe?

Feynman: Well, there's a lot of stories about safes. If you give me ten minutes, I'll tell you three stories about safes. All right? The motivation for me to open the filing cabinet, pick the lock, became my interest in the safety of the whole thing. Somebody had told me how to pick locks. Then they got filing cabinets which had safe combinations. One of my diseases, one of my things in life is that anything that is secret I try to undo. And so those locks to those filing cabinets, made by the Mosler Lock Company, in which we put our documents after that—everybody had them—they represented a challenge to me. How the hell to open them?! So I worked on them and I worked on them. There's all kinds of stories about how you can feel the numbers and listen to things and so on. That's true; I understand it, very well. For old-fashioned safes. They had a new design so that nothing would be pushing against the wheels while you were trying them. I won't go into the technical details, but none of the old methods would work. I read books by locksmiths. Books by locksmiths always say in the beginning how they opened the locks, the greatest thing, the woman is underwater, the safe is underwater and the woman is drowning or something, and he opened the safe. I don't know, crazy story. And then in the back they tell you how they do it and they don't tell you anything sensible; it doesn't sound like they could really open safes that way. Like *guess* the combination on the basis of the psychology of the person who owns it! So I always figured they'd keep it a secret. Anyway, I kept working. And so like a kind of disease, I kept working on these things until I found out a few things. First I found out how big a range you need to open the combination, how close you have to be. And then I invented a system by which you could try all the combinations that you have to try. Eight thousand as it turned out, because you could be within two of every number. Then it turns out that it's every fifth number out of a hundred and twenty thousand . . . eight thousand combinations. And then I worked out a scheme by which I could try numbers without altering a number that I already set, by correctly moving the wheels, so that I could do it in eight hours, try all the combinations. And then I discovered still further that—this took me about two years of researching—I had nothing to do up there, you see, and I was fiddling—finally I discovered a way that it's easy to take the numbers, the back two numbers, the last two numbers of the combination off the safe if the safe is open. If the drawer is pulled out you could turn the number and see the bolt go up and play around and find out what makes it, what number it comes back at and stuff like that. With a little trickery you can get the combination off. So I used to practice it like a cardsharp practices cards, you know, all the time, all the time. Quicker and quicker and more and more unobtrusively I would come in and I would talk to some guy and I'd sort of lean against his filing cabinet, just like I'm playing with this watch now; you wouldn't even notice I'm doing anything. I'm not doing anything. I would just play with the dial, that's all, just play with the dial. But I was taking the two numbers off! And I go back to my office and I write the two numbers down. The last two numbers of the three. Now, if you have the last two numbers, it takes a minute to try the first number; there's only twenty possibilities and it's open. OK?

So, I got an excellent reputation for safecracking. They would say to me, "Mr. Schmultz is out of town, we need a document from his safe. Can you open it?" I'd say, "Yes, I can open it; I have to go get my tools" (I don't need any tools). I go to my office and I'd look at the number of his safe. I had the last two numbers. I had everybody's safe numbers in my office. I put a screwdriver in my back pocket, to account for the tool I claimed I needed. I go back to the room and I would close the door. The attitude is that this business about how you open safes is not something that everybody should know because it makes everything very unsafe, it's very dangerous to have everybody know how to do this. So I close the door and then I sit down and I read a magazine, or do something. I'd average about 20 minutes of doing nothing, and then I'd open it, you see, well, I opened it right away to see that everything was all right and then I'd sit there for 20 minutes to give myself a good reputation that it wasn't too easy, there was no trick to it, no trick to it. And then I'd come out, you know, sweating a bit, and say "It's open. There you are," and so forth. OK?

Also, at one particular moment, I did open a safe purely by accident, and that helped to reinforce my reputation. It was a sensation, it was pure luck, the same kind of luck I had with the blueprints. But after the war was over, I've got to tell you these stories now because after the war was over I went back to Los Alamos to finish some papers and there I did some safe opening which—I could write a safecracker book *better* than any safecracker book. In the beginning it would explain how I opened the safe absolutely cold without knowing the combination, which contained *more* secret things than any safe that's ever been opened. I opened the safe that contained behind it the secret of the atomic bomb, *all* the secrets, the formulas, the rates at which neutrons are liberated from uranium, how much uranium you need to make a bomb, all the theories, all the calculations, the WHOLE DAMN THING!

This is the way it was done. All right? I was trying to write a report. I needed this report. It was a Saturday; I thought everybody worked. I thought it was like Los Alamos *used* to be. So I went down to get it from the library. The library at Los Alamos had all these documents. There was a great vault with a great knob of a different kind I didn't know anything about. Filing cabinets I understood, but I was only an expert on filing cabinets. Not only that but there were guards walking back and forth in front with guns. You can't get that one open, OK? But I think, wait! Old Freddy DeHoffman in the declassification section, he's in charge of declassifying documents. Which documents now can be declassified? And so he had to run down to the library and back so often, he got tired of it. And he got a brilliant idea. He would get a copy made of every document in the Los Alamos library. And he'd stick it in his file, he had *nine* filing cabinets, one right next to the other in two rooms, *full* of all the documents of Los Alamos and I knew he had that. So I'll go up to DeHoffman and I'll ask him to borrow the documents, from him, he's got a copy. So I went up to his office. The office door is open. It looks like he's coming back, the light is lit; looks like he's coming back any minute. So I wait. And as always when I'm waiting, I diddled the knobs. I tried 10-20-30, didn't work. I tried 20-40-60, didn't work. Tried everything. I'm waiting, nothing to do. Then I begin to think, you know, those locksmith people, I had never been able to figure out how to open them cleverly. Maybe they don't either, maybe all the stuff they're telling me about psychology is right. I'm going to open this one by psychology. First thing, the book says, "The secretary is very nervous that she will forget the combination." She's been told the combination. She might forget and the boss might forget—she has to know. So she nervously writes it somewhere. Where? List of places were a secretary might write combinations. OK? Starts out with, most clever thing, starts right out with—you open the drawer and on the wood along the side of the drawer, the outside, is written carelessly a number, like as if it is an invoice number. That's the combination number. So. It's on the side of the desk. OK? I remembered that, it's in the book. Desk drawer is locked, I picked the lock easy, I opened the lock right away, pull out the drawer, look along the wood—nothing. It's all right, it's all right. There's a lot of papers in the drawer. I fish around among the papers and finally I find it, a nice little piece of paper which has the Greek alphabet. Alpha, beta, gamma, delta, and so forth, carefully printed. The secretaries have to know how to make those letters and how to call them when they're talking about them, right? So they all had, each one had a copy of the thing. *But*–carelessly scrawled across the top is π is equal to 3.14159. Well, why does she need the numerical value of π, she's not computing anything? So I go up to the safe. Honest, it's honest, right? It's just like in the book. I'm just telling you how it was done. I walk up to the safe. 31-41-59. Doesn't open. 13-14-95–doesn't open. 95-14-13, doesn't open. 14-31, twenty minutes I'm turning π upside down. Nothing happens. So I start walking out of the office and I remember the book about the psychology and I said, you know, but it's true. Psychologically, I'm right. DeHoffman is *just* the kind of a guy to use a mathematical constant for his safe combination. So the other important mathematical constant is *e*. So I walk back to the safe, 27-18-28, click, clock, it opens. I checked by the way, that *all* the combinations were the same. Well, there's another lot of stories about it but it's getting late and that's a good one, so we'll let it go at that.

4

What Is and What
Should Be the Role of
Scientific Culture in
Modern Society

✦

Here is a talk Feynman gave to an audience of scientists at the Galileo Symposium in Italy, in 1964. With frequent acknowledgments and references to the great work and intense anguish of Galileo, Feynman speaks on the effect of science on religion, on society, and on philosophy, and warns that it is our capacity to doubt that will determine the future of civilization.

I am Professor Feynman, in spite of this suit-coat. I usually give lectures in shirtsleeves, but when I started out of the hotel this morning my wife said, "You must wear a suit." I said, "But I usually give lectures in shirtsleeves." She said, "Yes, but this time you don't know what you're talking about so you had better make a good impression. . . ." So, I got a coat.

The Pleasure of Finding Things Out

I am going to talk about the topic that was given me by Professor Bernardini.* I would like to say, at the very beginning, that, in my opinion, to find the proper place of scientific culture in modern society is not to solve the problems of modern society. There are a large number of problems that have nothing much to do with the position of science in society, and it is a dream to think that to simply decide on one aspect of how ideally science and society should be matched is somehow or other to solve all the problems. So, please understand that, although I will suggest some modifications of the relationship, I do not expect these modifications to be the solution to society's problems.

This modern society seems to be threatened by a number of serious threats, and the one that I would like to concentrate on and which will be in fact the central theme, although there will be a lot of subsidiary little items, the central theme of my discussion is that I believe that one of the greatest dangers to modern society is the possible resurgence and expansion of the ideas of thought control; such ideas as Hitler had, or Stalin in his time, or the Catholic religion in the Middle Ages, or the Chinese today. I think that one of the greatest dangers is that this shall increase until it encompasses all of the world.

Now, in discussing the relation of science to the scientific culture of society, the first thing that comes to mind immediately is, of course, the most obvious thing, which is the applications of science. The applications are culture, too. However, I am not going to talk about the applications, but not for any good reasons. I appreciate that all the popular discussions on the subject of the relation of science to society revolve around the applications almost completely, and furthermore that the moral questions that scientists have about

---

*Chairman of the conference. *Ed.*

the kind of work that they do also usually involve the applications. Nevertheless, I will not speak about them because there are a number of other items which are not spoken about by so many other people, and so for the fun of it I would like to talk in a slightly different direction.

I will, however, say about the applications that, as you all appreciate, science creates a power through its knowledge, a power to do things: You are able to do things after you know something scientifically. But the science does not give instructions with this power as to how to do good against how to do evil. Let us put it a very simple way: There are no instructions along with the power, and the question of applying the science or not is essentially the problem of organizing the applications in a way that doesn't do too much harm and does as much good as possible. But, of course, sometimes people in science try to say it is not their responsibility, because the application is just the power to do; it is independent of what you do with it. But it certainly is in some sense true that to create for mankind the power to control this is good, probably, in spite of the difficulties that he has in trying to figure out how to control the power to do himself good rather than evil.

May I say, too, that although many of us here are physicists, and most of us think of the serious problems of society in terms of physics, I believe most assuredly that the next science to find itself in moral difficulties with its applications is biology, and if the problems of physics relative to science seem difficult, the problems of the development of biological knowledge will be fantastic. These possibilities were hinted at, for example, in the book by Huxley, *Brave New World*, but you can think of a number of things. For example, if energy in the far future can be supplied freely and easily by physics, then it is a matter of mere chemistry to put together the atoms in such a way as to produce food, from energy that the atoms have conserved, so that you can produce as much food as there are waste products from human beings; and there is therefore a conservation of material and no food problem. There will be serious social problems when we find out how to control heredity, as to what kind of control, good or bad, to use. Suppose that we were to discover the physiological basis of happiness or other feelings, such as the feeling of ambition, and suppose that we could then control whether somebody feels ambitious or does not feel ambitious. And, finally, there is death.

It is one of the most remarkable things that in all of the biological sciences there is no clue as to the necessity of death. If you say we want to make perpetual motion, we have discovered enough laws as we studied physics to see that it is either absolutely impossible or else the laws are wrong. But there is nothing in biology yet found that indicates the inevitability of death. This suggests to me that it is not at all inevitable, and that it is only a matter of time before the biologists discover what it is that is causing us the trouble and that that terrible universal disease or temporariness of the human's body will be cured. Anyhow, you can see that there will be problems of a fantastic magnitude coming from biology.

Now I will talk in a different direction.

Besides the applications there are ideas, and the ideas are of two kinds. One of them is the product of the science itself, that is, a worldview which the science produces. This is in some ways the most beautiful part of the whole thing. Some people think, no, the methods of science are the thing. Well, it depends on whether you like the ends or the means, but the means were to produce some wonderful ends and I will not bore you (well, I wouldn't bore you if I could do it right) with the details. But you all know something about the wonders of science—it isn't a popular audience I am talking to—so I won't try to make you enthusiastic once again with the facts about the world: the fact that we are all made of atoms, the enormous ranges of time and space that there are, the position of ourselves historically as the result of a remarkable series of evolution. The position of ourselves in the evolutionary sequence; and further, the most remarkable aspect of our scientific worldview is its universality in this sense that although we talk about our being specialists, we are really not. One of the most promising hypotheses in all of biology is that everything the animals do or that living creatures do can be understood in terms of what atoms can do, that is, in terms of physical laws, ultimately, and the perpetual attention to this possibility—so far no exception has been demonstrated—has again and again made suggestions as to how the mechanisms actually occur. So that the fact that our knowledge is in fact universal is something that is not completely appreciated, that the position of the theories are so complete that we hunt for exceptions and we find them very hard to find—in the physics at least—and the great expense of all these machines and so on is to find some exception to what is already known. And, otherwise, that is another aspect of the fact that the world is so wonderful in the sense that stars are made of the same atoms as the cows and as ourselves, and as stones.

From time to time we all try to communicate to our unscientific friends this worldview—and we get into difficulty most often because we get confused in trying to explain to them the latest questions, such as the meaning of the conservation of CP,*

---

*Conservation of charge and parity, one of the fundamental conservation laws of physics, which says that the total electric charge and parity, an intrinsic symmetry property of subatomic particles, going into an interaction will be the same coming out of that interaction. *Ed.*

whereas they don't know anything about the most preliminary things. For four hundred years since Galileo we have
been gathering information about the world which they
don't know. Now we are working on something way out, and
at the limits of scientific knowledge. And the things that appear in the newspaper and that seem to excite the adult imagination are always those things which they cannot possibly
understand, because they haven't learned anything at all of
the much more interesting well-known [to scientists] things
that people have found out before. It's not the case with children, thank goodness, for a while—at least until they become
adults.

I say, and I think you must all know from experience, that
people—I mean the average person, the great majority of people, the enormous majority of people—are woefully, pitifully,
absolutely ignorant of the science of the world that they live
in, and they can stay that way. I don't mean to say the heck
with them, what I mean is that they are able to stay that way
without it worrying them at all—only mildly—so from time to
time when they see CP mentioned in the newspaper they ask
what it is. And an interesting question of the relation of science to modern society is just that—why is it possible for people to stay so woefully ignorant and yet reasonably happy in
modern society when so much knowledge is unavailable to
them?

Incidentally, about knowledge and wonder, Mr. Bernardini
said we shouldn't teach wonders but knowledge.

It may be a difference in the meaning of the words. I think
we should teach them wonders and that the purpose of
knowledge is to appreciate wonders even more. And that the
knowledge is just to put into correct framework the wonder
that nature is. However, he would probably agree that I just
shifted some words around and that meaning trickled into

the conversation. At any rate, I want to answer the question as to why people can remain so woefully ignorant and not get into difficulties in modern society. The answer is that science is irrelevant. And I will explain what I mean in just a minute. It isn't that it has to be, but that we let it be irrelevant to society. I will come back to that point.

The other aspects of science that are important and that have some problem of a relation to society, beside the applications and the actual facts that are discovered, are the ideas and the techniques of scientific investigation: the means, if you will. Because I think that it is hard to understand why the discovery of these means, which seem so self-evident and obvious, weren't discovered earlier; simple ideas which, if you just try them, you see what happens and so forth. It is probably that the human mind evolved from that of an animal; and it evolves in a certain way [such] that it is like any new tool, in that it has its diseases and difficulties. It has its troubles, and one of the troubles is that it gets polluted by its own superstitions, it confuses itself, and the discovery was finally made of a way to keep it sort of in line so that scientists can make a little progress in some direction rather than to go around in circles and force themselves into a hold. And I think that this is, of course, the appropriate time to discuss this matter because the beginnings of this new discovery were at the time of Galileo. These ideas and techniques, of course, you all know. I'll just review them; it's again one of those things that for a lay audience you go into great detail; I just mention them so you appreciate what I am talking about more specifically.

The first is the matter of judging evidence—well, the first thing really is, before you begin you must not know the answer. So you begin by being uncertain as to what the answer is. This is very, very important, so important that I would like to delay that aspect, and talk about that still further along in my speech. The question of doubt and uncertainty is what is necessary to begin; for if you already know the answer there is no need to gather any evidence about it. Well, being uncertain, the next thing is to look for evidence, and the scientific method is to begin with trials. But another way and a very important one that should not be neglected and that is very vital is to put together ideas to try to enforce a logical consistency among the various things that you know. It is a very valuable thing to try to connect this, what you know, with that, that you know, and try to find out if they are consistent. And the more activity in the direction of trying to put together the ideas of different directions, the better it is.

After we look for the evidence we have to judge the evidence. There are the usual rules about the judging the evidence; it's not right to pick only what you like, but to take all of the evidence, to try to maintain some objectivity about the thing—enough to keep the thing going—not to ultimately depend upon authority. Authority may be a hint as to what the truth is, but is not the source of information. As long as it's possible, we should disregard authority whenever the observations disagree with it. And finally, the recording of results should be done in a disinterested way. That's a very funny phrase which always bothers me—because it means that after the guy's all done with the thing, he doesn't give a darn about the results, but that isn't the point. Disinterest here means that they are not reported in such a way as to try to influence the reader into an idea that's different than what the evidence indicates.

And you all appreciate these various aspects.

Now all this, all these ideas, and all the techniques are in the spirit of Galileo. The man whose birthday we are celebrating had a great deal to do with the development and the spreading and, most importantly, the demonstration of the power of these ways of looking at things. In any centennial, or quattro-centennial likewise, one always gets the feeling sooner or later: I wonder if the man were here now and we showed him the world, what he would say. Of course, you say, that's a corny thing to do and you can't do that in a speech, but that's what I am going to do. Suppose Galileo were here and we were to show him the world today and try to make him happy, or see what he finds out. And we would tell him about the questions of evidence, those methods of judging things which he developed. And we would point out that we are still in exactly the same tradition, we follow it exactly—even to the details of making numerical measurements and using those as one of the better tools, in the physics at least. And that the sciences have developed in a very good way directly and continuously from his original ideas, in the same spirit he developed. And as a result there are no more witches and ghosts.

Actually I say [that the quantitative method works very well] in science, but that is in fact almost a definition of science today; the sciences that Galileo was worried about, the physics, mechanics and such things, have of course developed, but the same techniques worked in biology, in history, geology, anthropology, and so on. We know a great deal about the past history of man, the past history of animals, and of the earth, through very similar techniques. With somewhat similar success, but not quite as complete because of the difficulties, the same systems work in economics. But there are places where only lip service is paid to the forms—in which many people just go through the motions. I would be ashamed to tell Mr. Galileo, but it doesn't really work very well, for example, in the social sciences. For example, my own personal experience—as you will realize, there is an awful lot of studying of the methods of education going on, particularly of the teaching of arithmetic—but if you try to find out what is really known about what is the better way to teach arithmetic than some other way, you will discover that there is an enormous number of studies and a great deal of statistics, but they are all disconnected from one another and they are mixtures of anecdotes, uncontrolled experiments, and very poorly controlled experiments, so that there is very little information as a result.

And now finally, as I'd like to show Galileo our world, I must show him something with a great deal of shame. If we look away from the science and look at the world around us, we find out something rather pitiful: that the environment that we live in is so actively, intensely unscientific. Galileo could say: "I noticed that Jupiter was a ball with moons and not a god in the sky. Tell me, what happened to the astrologers?" Well, they print their results in the newspapers, in the United States at least, in every daily paper every day. Why do we still have astrologers? Why can someone write a book like *Worlds in Collision* by somebody with a name beginning with a "V," it's a Russian name? Huh? Vininkowski?* And how did it become popular? What is all this nonsense about Mary Brody, or something? I don't know, that was crazy stuff. There is always some crazy stuff. There is an infinite amount of crazy stuff, which, put another way, is that the environment is actively, intensely unscientific. There is talk about telepathy still, although it's dying out. There is faith-healing galore, all over. There is a whole religion of faith-healing. There's a miracle at Lourdes where healing goes on. Now, it might be true that as-

---

*Actually it was Immanuel Velikovsky: *Worlds in Collision* (Doubleday, New York, 1950). *Ed.*

trology is right. It might be true that if you go to the dentist on the day that Mars is at right angles to Venus, that it is better than if you go on a different day. It might be true that you can be cured by the miracle of Lourdes. But if it is true it ought to be investigated. Why? To improve it. If it is true then maybe we can find out if the stars do influence life; that we could make the system more powerful by investigating statistically, scientifically judging the evidence objectively, more carefully. If the healing process works at Lourdes, the question is how far from the site of the miracle can the person, who is ill, stand? Have they in fact made a mistake and the back row is really not working? Or is it working so well that there is plenty of room for more people to be arranged near the place of the miracle? Or is it possible, as it is with the saints which have recently been created in the United States—there is a saint who cured leukemia apparently indirectly—that ribbons that are touched to the sheet of the sick person (the ribbon having previously touched some relic of the saint) increase the cure of leukemia—the question is, is it gradually being diluted? You may laugh, but if you believe in the truth of the healing, then you are responsible to investigate it, to improve its efficiency and to make it satisfactory instead of cheating. For example, it may turn out that after a hundred touches it doesn't work anymore. Now it's also possible that the results of this investigation have other consequences, namely, that nothing is there.

And another thing that bothers me, I might as well mention, are the things that the theologians in modern times can discuss, without feeling ashamed of themselves. There are many things that they can discuss that they need not feel ashamed of themselves, but some of the things that go on in the conferences on religion, and the decisions that have to be made, are ridiculous in modern times. I would like to explain that one of the difficulties, and one of the reasons why this can keep going, is that it is not realized what a profound modification of our worldview would result, if just one example of one of these things would really work. The whole idea, if you could establish the truth, not of the whole idea of astrology but just one little item, could have a fantastic influence on our understanding of the world. And so the reasons we laugh a little bit is that we are so confident of our view of the world that we are sure they aren't going to contribute anything. On the other hand, why don't we get rid of it? I will come to why we don't get rid of it in a minute, because science is irrelevant [to astrology], as I said before.

Now I am going to mention still another thing which is a little more doubtful, but still I believe that in the judging of evidence, the reporting of evidence and so on, there is a kind of responsibility which the scientists feel toward each other which you can represent as a kind of morality. What's the right way and the wrong way to report results? Disinterestedly, so that the other man is free to understand precisely what you are saying, and as nearly as possible not covering it with your desires. That this is a useful thing, that this is a thing which helps each of us to understand each other, in fact to develop in a way that isn't personally in our own interest, but for the general development of ideas, is a very valuable thing. And so there is, if you will, a kind of scientific morality. I believe, hopelessly, that this morality should be extended much more widely; this idea, this kind of scientific morality, that such things as propaganda should be a dirty word. That a description of a country made by the people of another country should describe that country in a disinterested way. What a miracle—that's worse than a miracle at Lourdes! Advertising, for example, is an example of a scientifically immoral description of the products. This immorality is so extensive that one gets so used to it in ordinary life, that you do not appreciate that it is a bad thing. And I think that one of the important reasons to increase the contact of scientists with the rest of society is to explain, and to kind of wake them up to this permanent attrition of cleverness of the mind that comes from not having information, or [not] having information always in a form which is interesting.

There are other things in which scientific methods would be of some value; they are perfectly obvious but they get more and more difficult to discuss—such things as making decisions. I do not mean that it should be done scientifically, such as [the way] in the United States that the Rand Company sits down and makes arithmetical calculations. That reminds me of my sophomore days at college in which, in discussing women, we discovered that by using electrical terminology—impedance, reluctance, resistance—that we had a deeper understanding of the situation. The other thing that gives a scientific man the creeps in the world today are the methods of choosing leaders—in every nation. Today, for example, in the United States, the two political parties have decided to employ public relations men, that is, advertising men, who are trained in the necessary methods of telling the truth and lying in order to develop a product. This wasn't the original idea. They are supposed to discuss situations and not just make up slogans. It's true, if you look in history, however, that choosing political leaders in the United States has been on many different occasions based on slogans. (I am sure that each party now has million-dollar bank accounts and there are going to be some very clever slogans.) But I can't do a sum-up of all this stuff now.

I kept saying that the science was irrelevant. That sounds strange and I would like to come back to it. Of course it is relevant, because of the fact that it is relevant to astrology; because if we understand the world the way we do, we cannot understand how the astrological phenomena can take place. And so that is relevant. But for people who believe in astrology there is no relevance, because the scientist never bothers to argue with them. The people who believe in faith healing have not to worry about science at all, because nobody argues with them. You don't have to learn science if you don't feel like it. So you can forget the whole business if it is too much mental strain, which it usually is. Why can you forget the whole business? Because we don't do anything about it. I believe that we must attack these things in which we do not believe. Not attack by the method of cutting off the heads of the people, but attack in the sense of discuss. I believe that we should demand that people try in their own minds to obtain for themselves a more consistent picture of their own world; that they not permit themselves the luxury of having their brain cut in four pieces or two pieces even, and on one side they believe this and on the other side they believe that, but never try to compare the two points of view. Because we have learned that, by trying to put the points of view that we have in our head together and comparing one to the other, we make some progress in understanding and in appreciating where we are and what we are. And I believe that science has remained irrelevant because we wait until somebody asks us questions or until we are invited to give a speech on Einstein's theory to people who don't understand Newtonian mechanics, but we never are invited to give an attack on faith healing, or on astrology—on what is the scientific view of astrology today.

I think that we must mainly write some articles. Now what would happen? The person who believes in astrology will have to learn some astronomy. The person who believes in faith healing might have to learn some medicine, because of the arguments going back and forth; and some biology. In other words, it will be necessary that science become relevant. The remark which I read somewhere, that science is all right so long as it doesn't attack religion, was the clue that I needed to understand the problem. As long as it doesn't attack religion it need not be paid attention to and nobody has to learn anything. So it can be cut off from modern society except for its applications, and thus be isolated. And then we have this terrible struggle to try to explain things to people who have no reason to want to know. But if they want to defend their own point of view, they will have to learn what yours is a little bit. So I suggest, maybe incorrectly and perhaps wrongly, that we are too polite. There was in the past an era of conversation on these matters. It was felt by the church that Galileo's views attacked the church. It is not felt by the church today that the scientific views attack the church. Nobody is worrying about it. Nobody attacks; I mean, nobody writes trying to explain the inconsistencies between the theological views and the scientific views held by different people today—or even the inconsistencies sometimes held by the same scientist between his religious and scientific beliefs.

Now the next subject, and the last main subject that I want to talk about, is the one I really consider the most important and the most serious. And that has to do with the question of uncertainty and doubt. A scientist is never certain. We all know that. We know that all our statements are approximate statements with different degrees of certainty; that when a statement is made, the question is not whether it is true or false but rather how likely it is to be true or false. "Does God exist?" "When put in the questional form, how likely is it?" It makes such a terrifying transformation of the religious point of view, and that is why the religious point of view is unscientific. We must discuss each question within the uncertainties that are allowed. And as evidence grows it increases the probability perhaps that some idea is right, or decreases it. But it never makes absolutely certain one way or the other. Now we have found that this is of paramount importance in order to progress. We absolutely must leave room for doubt or there is no progress and there is no learning. There is no learning without having to pose a question. And a question requires doubt. People search for certainty. But there *is* no certainty. People are terrified—how can you live *and not know*? It is not odd at all. You only think you know, as a matter of fact. And most of your actions are based on incomplete knowledge and you really don't know what it is all about, or what the purpose of the world is, or know a great deal of other things. It is possible to live and not know.

Now the freedom to doubt, which is absolutely essential for the development of the sciences, was born from a struggle with the constituted authorities of the time who had a solution to every problem, namely, the church. Galileo is a symbol of that struggle—one of the most important strugglers. And although Galileo himself apparently was forced to recant, nobody takes the confession seriously. We do not feel that we should follow Galileo in this way and that we should all recant. In fact, we consider the recantation as a foolishness—that the church asked for such a foolishness that we see again and again; and we feel sympathetic to Galileo as we feel sympathetic to the musicians and the artists of the Soviet Union who had to recant, and fortunately in apparently somewhat fewer numbers in recent times. But the recantation is a meaningless thing, no matter how cleverly it is organized. It is perfectly obvious to people from the outside that it is nothing to consider, and that Galileo's recantation is not something that we need to discuss as demonstrating anything about Galileo, except that perhaps he was an old man and that the church was very powerful. The fact that Galileo was right is not essential to this discussion. The fact that he was trying to be suppressed is, of course.

We are all saddened when we look at the world and see what few accomplishments we have made, compared to what we feel are the potentialities of human beings. People in the past, in the nightmare of their times, had dreams for the future. And now that the future has materialized we see that in many ways the dreams have been surpassed, but in still more ways many of our dreams of today are very much the dreams of people of the past. There have, in the past, been great enthusiasms for one or another's method of solving a problem. One was that education should become universal, for then all men would become Voltaires, and then we would have everything straightened out. Universal education is probably a good thing, but you could teach bad as well as good—you [could] teach falsehood as well as truth. The communication between nations as it develops through a technical development of science should certainly improve the relations between nations. Well, it depends what you communicate. You can communicate truth and you can communicate lies. You can communicate threats or kindnesses. There was a great hope that the applied sciences would free man of his physical struggles, and particularly in medicine it seems, for example, that all is to the good. Yes, but while we are talking, scientists are working in hidden secret laboratories trying to develop, as best they can, diseases which the other man can't cure. Perhaps today we have the dream that economic satiation of all men is the solution to the problem. I mean everybody should have enough stuff. I don't mean, of course, that we shouldn't try to do that. I don't mean, by what I'm saying, that we should not educate, or that we should not communicate, or that we shouldn't produce economic satiation.

But that this is the solution all by itself, of all problems, is doubtful. Because in those places where we have a certain degree of economic satisfaction, we have a whole host of new problems, or probably old problems that just look a little different because we happen to know enough about history.

So today we are not very well off, we don't see that we have done too well. Men, philosophers of all ages, have tried to find the secret of existence, the meaning of it all. Because if they could find the real meaning of life, then all this human effort, all this wonderful potentiality of human beings, could then be moved in the correct direction and we would march forward with great success. So therefore we tried these different ideas. But the question of the meaning of the whole world, of life, and of human beings, and so on, has been answered very many times by very many people. Unfortunately all the answers are different; and the people with one answer look with horror at the actions and behavior of the people with another answer. Horror, because they see the terrible things that are done; the way man is being pushed into a blind alley by this rigid view of the meaning of the world. In fact, it is really perhaps by the fantastic size of the horror that it becomes clear how great are the potentialities of human beings, and it is possibly this which makes us hope that if we could move things in the right direction, things would be much better.

What then is the meaning of the whole world? We do not know what the meaning of existence is. We say, as the result of studying all of the views that we have had before, we find that we do not know the meaning of existence; but in saying that we do not know the meaning of existence, we have probably found the open channel—if we will allow only that, as we progress, we leave open opportunities for alternatives, that we do not become enthusiastic for the fact, the knowledge, the absolute truth, but remain always uncertain—[that we] "hazard it." The English, who have developed their government in this direction, call it "muddling through," and although a rather silly, stupid sounding thing, it is the most scientific way of progressing. To decide upon the answer is not scientific. In order to make progress, one must leave the door to the unknown ajar—ajar only. We are only at the beginning of the development of the human race; of the development of the human mind, of intelligent life—we have years and years in the future. It is our responsibility not to give the answer today as to what it is all about, to drive everybody down in that direction and to say: "This is a solution to it all." Because we will be chained then to the limits of our present imagination. We will only be able to do those things that we think today are the things to do. Whereas, if we leave always some room for doubt, some room for discussion, and proceed in a way analogous to the sciences, then this difficulty will not arise. I believe, therefore, that although it is not the case today, that there may some day come a time, I should hope, when it will be fully appreciated that the power of government should be limited; that governments ought not to be empowered to decide the validity of scientific theories, that that is a ridiculous thing for them to try to do; that they are not to decide the various descriptions of history or of economic theory or of philosophy. Only in this way can the real possibilities of the future human race be ultimately developed.

5

There's Plenty of Room at the Bottom

✦

In this famous talk to the American Physical Society on December 29, 1959, at Caltech, Feynman, the "father of nanotechnology," expounds, decades ahead of his time, on the future of miniaturization: how to put the entire Encyclopaedia Brittanica on the head of a pin, the drastic reduction in size of both biological and inanimate objects, and the problems of lubricating machines smaller than the period at the end of this sentence. Feynman makes his famous wager, challenging young scientists to construct a working motor no more than 1/64 of an inch on all sides.

An Invitation to Enter a New Field of Physics

I imagine experimental physicists must often look with envy at men like Kamerlingh-Onnes,* who discovered a field like low temperature, which seems to be bottomless and in which

---

*Heike Kamerlingh-Onnes (1853–1926), winner of the 1913 Physics Nobel Prize for investigations of the properties of matter at low temperatures, which led to the production of liquid helium. *Ed.*

one can go down and down. Such a man is then a leader and has some temporary monopoly in a scientific adventure. Percy Bridgman,* in designing a way to obtain higher pressures, opened up another new field and was able to move into it and to lead us all along. The development of ever higher vacuum was a continuing development of the same kind.

I would like to describe a field in which little has been done, but in which an enormous amount can be done in principle. This field is not quite the same as the others in that it will not tell us much of fundamental physics (in the sense of, "What are the strange particles?"), but it is more like solid-state physics in the sense that it might tell us much of great interest about the strange phenomena that occur in complex situations. Furthermore, a point that is most important is that it would have an enormous number of technical applications.

What I want to talk about is the problem of manipulating and controlling things on a small scale.

As soon as I mention this, people tell me about miniaturization, and how far it has progressed today. They tell me about electric motors that are the size of the nail on your small finger. And there is a device on the market, they tell me, by which you can write the Lord's Prayer on the head of a pin. But that's nothing; that's the most primitive, halting step in the direction I intend to discuss. It is a staggeringly small world that is below. In the year 2000, when they look back at this age, they will wonder why it was not until the year 1960 that anybody began seriously to move in this direction.

Why cannot we write the entire 24 volumes of the Encyclopaedia Brittanica *on the head of a pin?*

---

*(1882–1961) Winner of the 1946 Physics Nobel Prize for his invention of an apparatus for producing extremely high pressures, and further work in high pressure physics. *Ed.*

There's Plenty of Room at the Bottom

Let's see what would be involved. The head of a pin is a six-teenth of an inch across. If you magnify it by 25,000 diameters, the area of the head of the pin is then equal to the area of all the pages of the *Encyclopaedia Brittanica*. Therefore, all it is necessary to do is to reduce in size all the writing in the *Encyclopaedia* by 25,000 times. Is that possible? The resolving power of the eye is about 1/120 of an inch—that is roughly the diameter of one of the little dots on the fine half-tone reproductions in the *Encyclopaedia*. This, when you demagnify it by 25,000 times, is still 80 angstroms* in diameter—32 atoms across, in an ordinary metal. In other words, one of those dots still would contain in its area 1,000 atoms. So, each dot can easily be adjusted in size as required by the photoengraving, and there is no question that there is enough room on the head of a pin to put all of the *Encyclopaedia Brittanica*.

Furthermore, it can be read if it is so written. Let's imagine that it is written in raised letters of metal; that is, where the black is in the *Encyclopaedia,* we have raised letters of metal that are actually 1/25,000 of their ordinary size. How would we read it?

If we had something written in such a way, we could read it using techniques in common use today. (They will un-doubtedly find a better way when we do actually have it writ-ten, but to make my point conservatively I shall just take techniques we know today.) We would press the metal into a plastic material and make a mold of it, then peel the plastic off very carefully, evaporate silica into the plastic to get a very thin film, then shadow it by evaporating gold at an angle against the silica so that all the little letters will appear clearly, dissolve the plastic away from the silica film, and then look through it with an electron microscope!

---

*One angstrom = one ten-billionth of a meter. *Ed.*

The Pleasure of Finding Things Out

There is no question that if the thing were reduced by 25,000 times in the form of raised letters on the pin, it would be easy for us to read it today. Furthermore, there is no question that we would find it easy to make copies of the master; we would just need to press the same metal plate again into plastic and we would have another copy.

How Do We Write Small?

The next question is: How do we *write* it? We have no standard technique to do this now. But let's argue that it is not as difficult as it first appears to be. We can reverse the lenses of the electron microscope in order to demagnify as well as magnify. A source of ions, sent through the microscope lenses in reverse, could be focused to a very small spot. We could write with that spot like we write in a TV cathode ray oscilloscope, by going across in lines, and having an adjustment which determines the amount of material which is going to be deposited as we scan in lines.

This method might be very slow because of space charge limitations. There will be more rapid methods. We could first make, perhaps by some photo process, a screen which has holes in it in the form of the letters. Then we would strike an arc behind the holes and draw metallic ions through the holes; then we could again use our system of lenses and make a small image in the form of ions, which would deposit the metal on the pin.

A simpler way might be this (though I am not sure it would work): We take light and, through an optical microscope running backwards, we focus it onto a very small photoelectric screen. Then electrons come away from the screen where the light is shining. These electrons are focused down in size by the electron microscope lenses to impinge directly upon the surface of the metal. Will such a beam etch away the metal if it is run long enough? I don't know. If it doesn't work for a metal surface, it must be possible to find some surface with which to coat the original pin so that, where the electrons bombard, a change is made which we could recognize later.

There is no intensity problem in these devices—not what you are used to in magnification, where you have to take a few electrons and spread them over a bigger and bigger screen; it is just the opposite. The light which we get from a page is concentrated onto a very small area so it is very intense. The few electrons which come from the photoelectric screen are demagnified down to a very tiny area so that, again, they are very intense. I don't know why this hasn't been done yet!

That's the *Encyclopaedia Brittanica* on the head of a pin, but let's consider all the books in the world. The Library of Congress has approximately 9 million volumes; the British Museum Library has 5 million volumes; there are also 5 million volumes in the National Library in France. Undoubtedly there are duplications, so let us say that there are some 24 million volumes of interest in the world.

What would happen if I print all this down at the scale we have been discussing? How much space would it take? It would take, of course, the area of about a million pinheads because, instead of there being just the 24 volumes of the *Encyclopaedia*, there are 24 million volumes. The million pinheads can be put in a square of a thousand pins on a side, or an area of about 3 square yards. That is to say, the silica replica with the paper-thin backing of plastic, with which we have made the copies, with all this information, is on an area of approximately the size of 35 pages of the *Encyclopaedia*. That is about half as many pages as there are in this magazine. All of the information which all of mankind has ever recorded in books can be carried around in a pamphlet in your hand—and not written in code, but a simple reproduction of the original pictures, engravings, and everything else on a small scale without loss of resolution.

What would our librarian at Caltech say, as she runs all over from one building to another, if I tell her that, ten years from now, all of the information that she is struggling to keep track of—120,000 volumes, stacked from the floor to the ceiling, drawers full of cards, storage rooms full of the older books—can be kept on just one library card! When the University of Brazil, for example, finds that their library is burned, we can send them a copy of every book in our library by striking off a copy from the master plate in a few hours and mailing it in an envelope no bigger or heavier than any other ordinary air mail letter.

Now, the name of this talk is "There Is *Plenty* of Room at the Bottom"—not just "There Is Room at the Bottom." What I have demonstrated is that there is room—that you can decrease the size of things in a practical way. I now want to show that there is *plenty* of room. I will not now discuss how we are going to do it, but only what is possible in principle—in other words, what is possible according to the laws of physics. I am not inventing anti-gravity, which is possible someday only if the laws are not what we think. I am telling you what could be done if the laws *are* what we think; we are not doing it simply because we haven't yet gotten around to it.

Information on a Small Scale

Suppose that, instead of trying to reproduce the pictures and all the information directly in its present form, we write only the information content in a code of dots and dashes, or something like that, to represent the various letters. Each letter represents six or seven "bits" of information; that is, you need only about six or seven dots or dashes for each letter. Now, instead of writing everything, as I did before, on the *surface* of the head of a pin, I am going to use the interior of the material as well.

Let us represent a dot by a small spot of one metal, the next dash by an adjacent spot of another metal, and so on. Suppose, to be conservative, that a bit of information is going to require a little cube of atoms 5 times 5 times 5—that is, 125 atoms. Perhaps we need a hundred and some odd atoms to make sure that the information is not lost through diffusion, or through some other process.

I have estimated how many letters there are in the *Encyclopaedia*, and I have assumed that each of my 24 million books is as big as an *Encyclopaedia* volume, and have calculated, then, how many bits of information there are (10^{15}). For each bit I allow 100 atoms. And it turns out that all of the information that man has carefully accumulated in all the books in the world can be written in this form in a cube of material one two-hundredths of an inch wide—which is the barest piece of dust that can be made out by the human eye. So there is *plenty* of room at the bottom! Don't tell me about microfilm!

This fact—that enormous amounts of information can be carried in an exceedingly small space—is, of course, well known to the biologists, and resolves the mystery which existed before we understood all this clearly, of how it could be that, in the tiniest cell, all of the information for the organization of a complex creature such as ourselves can be stored. All this information—whether we have brown eyes, or whether we think at all, or that in the embryo the jawbone should first develop with a little hole in the side so that later a nerve can grow through it—all this information is contained in a very tiny fraction of the cell in the form of long-chain

The Pleasure of Finding Things Out

DNA molecules in which approximately 50 atoms are used for one bit of information about the cell.

Better Electron Microscopes

If I have written in a code, with 5 times 5 times 5 atoms to a bit, the question is: How could I read it today? The electron microscope is not quite good enough; with the greatest care and effort, it can only resolve about 10 angstroms. I would like to try and impress upon you while I am talking about all of these things on a small scale, the importance of improving the electron microscope by a hundred times. It is not impossible; it is not against the laws of diffraction of the electron. The wave length of the electron in such a microscope is only 1/20 of an angstrom. So it should be possible to see the individual atoms. What good would it be to see individual atoms distinctly?

We have friends in other fields—in biology, for instance. We physicists often look at them and say, "You know the reason you fellows are making so little progress?" (Actually I don't know any field where they are making more rapid progress than they are in biology today.) "You should use more mathematics, like we do." They could answer us—but they're polite, so I'll answer for them: "What you should do in order for *us* to make more rapid progress is to make the electron microscope 100 times better."

What are the most central and fundamental problems of biology today? They are questions like: What is the sequence of bases in the DNA? What happens when you have a mutation? How is the base order in the DNA connected to the order of amino acids in the protein? What is the structure of the RNA; is it single-chain or double-chain, and how is it related in its order of bases to the DNA? What is the organization of the microsomes? How are proteins synthesized? Where does the RNA go? How does it sit? Where do the proteins sit? Where do the amino acids go in? In photosynthesis, where is the chlorophyll; how is it arranged; where are the carotenoids involved in this thing? What is the system of the conversion of light into chemical energy?

It is very easy to answer many of these fundamental biological questions; you just *look at the thing*! You will see the order of bases in the chain; you will see the structure of the microsome. Unfortunately, the present microscope sees at a scale which is just a bit too crude. Make the microscope one hundred times more powerful, and many problems of biology would be made very much easier. I exaggerate, of course, but the biologists would surely be very thankful to you—and they would prefer that to the criticism that they should use more mathematics.

The theory of chemical processes today is based on theoretical physics. In this sense, physics supplies the foundation of chemistry. But chemistry also has analysis. If you have a strange substance and you want to know what it is, you go through a long and complicated process of chemical analysis. You can analyze almost anything today, so I am a little late with my idea. But if the physicists wanted to, they could also dig under the chemists in the problem of chemical analysis. It would be very easy to make an analysis of any complicated chemical substance; all one would have to do would be to look at it and see where the atoms are. The only trouble is that the electron microscope is one hundred times too poor. (Later, I would like to ask the question: Can the physicists do something about the third problem of chemistry—namely, synthesis? Is there a *physical* way to synthesize any chemical substance?)

The reason the electron microscope is so poor is that the f-value of the lenses is only 1 part to 1,000; you don't have a big enough numerical aperture. And I know that there are theorems which prove that it is impossible, with axially symmetrical stationary field lenses, to produce an f-value any bigger than so and so; and therefore the resolving power at the present time is at its theoretical maximum. But in every theorem there are assumptions. Why must the field be symmetrical? I put this out as a challenge: Is there no way to make the electron microscope more powerful?

The Marvelous Biological System

The biological example of writing information on a small scale has inspired me to think of something that should be possible. Biology is not simply writing information; it is *doing something* about it. A biological system can be exceedingly small. Many of the cells are very tiny, but they are very active; they manufacture various substances; they walk around; they wiggle; and they do all kinds of marvelous things—all on a very small scale. Also, they store information. Consider the possibility that we too can make a thing very small which does what we want—that we can manufacture an object that maneuvers at that level!

There may even be an economic point to this business of making things very small. Let me remind you of some of the problems of computing machines. In computers we have to store an enormous amount of information. The kind of writing that I was mentioning before, in which I had everything down as a distribution of metal, is permanent. Much more interesting to a computer is a way of writing, erasing, and writing something else. (This is usually because we don't want to waste the material on which we have just written. Yet if we could write it in a very small space, it wouldn't make any difference; it could just be thrown away after it was read. It doesn't cost very much for the material.)

Miniaturizing the Computer

I don't know how to do this on a small scale in a practical way, but I do know that computing machines are very large; they fill rooms. Why can't we make them very small, make them of little wires, little elements—and by little, I mean *little*. For instance, the wires should be 10 or 100 atoms in diameter, and the circuits should be a few thousand angstroms across. Everybody who has analyzed the logical theory of computers has come to the conclusion that the possibilities of computers are very interesting—if they could be made to be more complicated by several orders of magnitude. If they had millions of times as many elements, they could make judgments. They would have time to calculate what is the best way to make the calculation that they are about to make. They could select the method of analysis which, from their experience, is better than the one that we would give to them. And in many other ways, they would have new qualitative features.

If I look at your face I immediately recognize that I have seen it before. (Actually, my friends will say I have chosen an unfortunate example here for the subject of this illustration. At least I recognize that it is a *man* and not an *apple*.) Yet there is no machine which, with that speed, can take a picture of a face and say even that it is a man; and much less that it is the same man that you showed it before—unless it is exactly the same picture. If the face is changed; if I am closer to the face; if I am further from the face; if the light changes—I recognize it anyway. Now, this little computer I carry in my head is easily able to do that. The computers that we build are not able to do that. The number of elements in this bone box of mine are enormously greater than the number of elements in our "wonderful" computers. But our mechanical computers are too big; the elements in this box are microscopic. I want to make some that are *sub*microscopic.

If we wanted to make a computer that had all these marvelous extra qualitative abilities, we would have to make it, perhaps, the size of the Pentagon. This has several disadvantages. First, it requires too much material; there may not be enough germanium in the world for all the transistors which would have to be put into this enormous thing. There is also the problem of heat generation and power consumption; TVA would be needed to run the computer. But an even more practical difficulty is that the computer would be limited to a certain speed. Because of its large size, there is finite time required to get the information from one place to another. The information cannot go any faster than the speed of light—so, ultimately, when our computers get faster and faster and more and more elaborate, we will have to make them smaller and smaller.

But there is plenty of room to make them smaller. There is nothing that I can see in the physical laws that says the computer elements cannot be made enormously smaller than they are now. In fact, there may be certain advantages.

Miniaturization by Evaporation

How can we make such a device? What kind of manufacturing processes would we use? One possibility we might consider, since we have talked about writing by putting atoms down in a certain arrangement, would be to evaporate the material, then evaporate the insulator next to it. Then, for the next layer, evaporate another position of a wire, another insulator, and so on. So, you simply evaporate until you have a block of stuff which has the elements—coils and condensers, transistors and so on—of exceedingly fine dimensions.

But I would like to discuss, just for amusement, that there are other possibilities. Why can't we manufacture these small computers somewhat like we manufacture the big ones? Why can't we drill holes, cut things, solder things, stamp things out, mold different shapes all at an infinitesimal level? What are the limitations as to how small a thing has to be before you can no longer mold it? How many times when you are working on something frustratingly tiny, like your wife's wrist-watch, have you said to yourself, "If I could only train an ant to do this!" What I would like to suggest is the possibility of training an ant to train a mite to do this. What are the possibilities of small but movable machines? They may or may not be useful, but they surely would be fun to make.

Consider any machine—for example, an automobile—and ask about the problems of making an infinitesimal machine like it. Suppose, in the particular design of the automobile, we need a certain precision of the parts; we need an accuracy, let's suppose, of 4/10,000 of an inch. If things are more inaccurate than that in the shape of the cylinder and so on, it isn't going to work very well. If I make the thing too small, I have to worry about the size of the atoms; I can't make a circle of "balls" so to speak, if the circle is too small. So, if I make the error, corresponding to 4/10,000 of an inch, correspond to an error of 10 atoms, it turns out that I can reduce the dimensions of an automobile 4,000 times, approximately—so that it is 1 mm across. Obviously, if you redesign the car so that it would work with a much larger tolerance, which is not at all impossible, then you could make a much smaller device.

It is interesting to consider what the problems are in such small machines. Firstly, with parts stressed to the same degree, the forces go as the area you are reducing, so that things like weight and inertia are of relatively no importance. The strength of material, in other words, is very much greater in proportion. The stresses and expansion of the flywheel from centrifugal force, for example, would be the same proportion only if the rotational speed is increased in the same proportion as we decrease the size. On the other hand, the metals that we use have a grain structure, and this would be very annoying at small scale because the material is not homogeneous. Plastics and glass and things of this amorphous nature are very much more homogeneous, and so we would have to make our machines out of such materials.

There are problems associated with the electrical part of the system—with the copper wires and the magnetic parts. The magnetic properties on a very small scale are not the same as on a large scale; there is the "domain" problem involved. A big magnet made of millions of domains can only be made on a small scale with one domain. The electrical equipment won't simply be scaled down; it has to be redesigned. But I can see no reason why it can't be redesigned to work again.

Problems of Lubrication

Lubrication involves some interesting points. The effective viscosity of oil would be higher and higher in proportion as we went down (and if we increase the speed as much as we can). If we don't increase the speed so much, and change from oil to kerosene or some other fluid, the problem is not so bad. But actually we may not have to lubricate at all! We have a lot of extra force. Let the bearings run dry; they won't run hot because the heat escapes away from such a small device very, very rapidly. This rapid heat loss would prevent the gasoline from exploding, so an internal combustion engine is impossible. Other chemical reactions, liberating energy when cold, can be used. Probably an external supply of electrical power would be most convenient for such small machines.

What would be the utility of such machines? Who knows? Of course, a small automobile would only be useful for the mites to drive around in, and I suppose our Christian interests don't go that far. However, we did note the possibility of the manufacture of small elements for computers in completely automatic factories, containing lathes and other machine tools at the very small level. The small lathe would not have to be exactly like our big lathe. I leave to your imagination the improvement of the design to take full advantage of the properties of things on a small scale, and in such a way that the fully automatic aspect would be easiest to manage.

A friend of mine (Albert R. Hibbs)* suggests a very interesting possibility for relatively small machines. He says that, although it is a very wild idea, it would be interesting in surgery if you could swallow the surgeon. You put the mechanical surgeon inside the blood vessel and it goes into the heart and "looks" around. (Of course the information has to be fed out.) It finds out which valve is the faulty one and takes a little knife and slices it out. Other small machines might be permanently incorporated in the body to assist some inadequately-functioning organ.

Now comes the interesting question: How do we make such a tiny mechanism? I leave that to you. However, let me suggest one weird possibility. You know, in the atomic energy

---

*A student and later a colleague of Feynman. *Ed.*

plants they have materials and machines that they can't handle directly because they have become radioactive. To unscrew nuts and put on bolts and so on, they have a set of master and slave hands, so that by operating a set of levers here, you control the "hands" there, and can turn them this way and that so you can handle things quite nicely.

Most of these devices are actually made rather simply, in that there is a particular cable, like a marionette string, that goes directly from the controls to the "hands." But, of course, things also have been made using servo motors, so that the connection between the one thing and the other is electrical rather than mechanical. When you turn the levers, they turn a servo motor, and it changes the electrical currents in the wires, which repositions a motor at the other end.

Now, I want to build much the same device—a master-slave system which operates electrically. But I want the slaves to be made especially carefully by modern large-scale machinists so that they are one-fourth the scale of the "hands" that you ordinarily maneuver. So you have a scheme by which you can do things at one-quarter scale anyway—the little servo motors with little hands play with little nuts and bolts; they drill little holes; they are four times smaller. Aha! So I manufacture a quarter-size lathe; I manufacture quarter-size tools; and I make, at the one-quarter scale, still another set of hands again relatively one-quarter size! This is one-sixteenth size, from my point of view. And after I finish doing this I wire directly from my large-scale system, through transformers perhaps, to the one-sixteenth-size servo motors. Thus I can now manipulate the one-sixteenth-size hands.

Well, you get the principle from there on. It is rather a difficult program, but it is a possibility. You might say that one can go much farther in one step than from one to four. Of course, this has all to be designed very carefully and it is not necessary simply to make it like hands. If you thought of it very carefully, you could probably arrive at a much better system for doing such things.

If you work through a pantograph, even today, you can get much more than a factor of four in even one step. But you can't work directly through a pantograph which makes a smaller pantograph which then makes a smaller pantograph—because of the looseness of the holes and the irregularities of construction. The end of the pantograph wiggles with a relatively greater irregularity than the irregularity with which you move your hands. In going down this scale, I would find the end of the pantograph on the end of the pantograph on the end of the pantograph shaking so badly that it wasn't doing anything sensible at all.

At each stage, it is necessary to improve the precision of the apparatus. If, for instance, having made a small lathe with a pantograph, we find its lead screw irregular—more irregular than the large-scale one—we could lap the lead screw against breakable nuts that you can reverse in the usual way back and forth until this lead screw is, at its scale, as accurate as our original lead screws, at our scale.

We can make flats by rubbing unflat surfaces in triplicates together—in three pairs—and the flats then become flatter than the thing you started with. Thus, it is not impossible to improve precision on a small scale by the correct operations. So, when we build this stuff, it is necessary at each step to improve the accuracy of the equipment by working for a while down there, making accurate lead screws, Johansen blocks, and all the other materials which we use in accurate machine work at the higher level. We have to stop at each level and manufacture all the stuff to go to the next level—a very long and very difficult program. Perhaps you can figure a better way than that to get down to small scale more rapidly.

Yet, after all this, you have just got one little baby lathe four thousand times smaller than usual. But we were thinking of making an enormous computer, which we were going to build by drilling holes on this lathe to make little washers for the computer. How many washers can you manufacture on this one lathe?

A Hundred Tiny Hands

When I make my first set of slave "hands" at one-fourth scale, I am going to make ten sets. I make ten sets of "hands," and I wire them to my original levers so they each do exactly the same thing at the same time in parallel. Now, when I am making my new devices one-quarter again as small, I let each one manufacture ten copies, so that I would have a hundred "hands" at the 1/16 size.

Where am I going to put the million lathes that I am going to have? Why, there is nothing to it; the volume is much less than that of even one full-scale lathe. For instance, if I made a billion little lathes, each 1/4000 of the scale of a regular lathe, there are plenty of materials and space available because in the billion little ones there is less than 2 percent of the materials in one big lathe. It doesn't cost anything for materials, you see. So I want to build a billion tiny factories, models of each other, which are manufacturing simultaneously, drilling holes, stamping parts, and so on.

As we go down in size, there are a number of interesting problems that arise. All things do not simply scale down in proportion. There is the problem that materials stick together by the molecular (Van der Waals*) attractions. It would be

---

*Van der Waals forces: Weak attractive forces between atoms or molecules. Johannes Diderik Van der Waals (1837–1923) received the 1910 Nobel Prize in Physics for his work on the equation of state for gases and liquids. *Ed.*

like this: After you have made a part and you unscrew the nut from a bolt, it isn't going to fall down because the gravity isn't appreciable; it would even be hard to get it off the bolt. It would be like those old movies of a man with his hands full of molasses, trying to get rid of a glass of water. There will be several problems of this nature that we will have to be ready to design for.

Rearranging the Atoms

But I am not afraid to consider the final question as to whether, ultimately—in the great future—we can arrange the atoms the way we want; the very *atoms*, all the way down! What would happen if we could arrange the atoms one by one the way we want them (within reason, of course; you can't put them so that they are chemically unstable, for example)?

Up to now, we have been content to dig in the ground to find minerals. We heat them and we do things on a large scale with them, and we hope to get a pure substance with just so much impurity, and so on. But we must always accept some atomic arrangement that nature gives us. We haven't got anything, say, with a "checkerboard" arrangement, with the impurity atoms exactly arranged 1,000 angstroms apart, or in some other particular pattern.

What could we do with layered structures with just the right layers? What would the properties of materials be if we could really arrange the atoms the way we want them? They would be very interesting to investigate theoretically. I can't see exactly what would happen, but I can hardly doubt that when we have some *control* of the arrangement of things on a small scale, we will get an enormously greater range of possible properties that substances can have, and of different things that we can do.

Consider, for example, a piece of material in which we make little coils and condensers (or their solid-state analogs) 1,000 or 10,000 angstroms in a circuit, one right next to the other, over a large area, with little antennas sticking out at the other end—a whole series of circuits. Is it possible, for example, to emit light from a whole set of antennas, like we emit radio waves from an organized set of antennas to beam the radio programs to Europe? The same thing would be to beam light out in a definite direction with very high intensity. (Perhaps such a beam is not very useful technically or economically.)

I have thought about some of the problems of building electric circuits on a small scale, and the problem of resistance is serious. If you build a corresponding circuit on a small scale, its natural frequency goes up, since the wave length goes down as the scale; but the skin depth only decreases with the square root of the scale ratio, and so resistive problems are of increasing difficulty. Possibly we can beat resistance through the use of superconductivity if the frequency is not too high, or by other tricks.

Atoms in a Small World

When we get to the very, very small world—say, circuits of seven atoms—we have a lot of new things that would happen that represent completely new opportunities for design. Atoms on a small scale behave like *nothing* on a large scale, for they satisfy the laws of quantum mechanics. So, as we go down and fiddle around with the atoms down there, we are working with different laws, and we can expect to do different things. We can manufacture in different ways. We can use, not just circuits, but some system involving the quantized energy levels, or the interactions of quantized spins, etc.

Another thing we will notice is that, if we go down far enough, all of our devices can be mass produced so that they are absolutely perfect copies of one another. We cannot build two large machines so that the dimensions are exactly the same. But if your machine is only 100 atoms high, you only have to get it correct to one-half of one percent to make sure the other machine is exactly the same size—namely, 100 atoms high!

At the atomic level, we have new kinds of forces and new kinds of possibilities, new kinds of effects. The problems of manufacture and reproduction of materials will be quite different. I am, as I said, inspired by the biological phenomena in which chemical forces are used in repetitious fashion to produce all kinds of weird effects (one of which is the author). The principles of physics, as far as I can see, do not speak against the possibility of maneuvering things atom by atom. It is not an attempt to violate any laws; it is something, in principle, that can be done; but in practice, it has not been done because we are too big.

Ultimately, we can do chemical synthesis. A chemist comes to us and says, "Look, I want a molecule that has the atoms arranged thus and so; make me that molecule." The chemist does a mysterious thing when he wants to make a molecule. He sees that it has got that ring, so he mixes this and that, and he shakes it, and he fiddles around. And, at the end of a difficult process, he usually does succeed in synthesizing what he wants. By the time I get my devices working, so that we can do it by physics, he will have figured out how to synthesize absolutely anything, so that this will really be useless.

But it is interesting that it would be, in principle, possible (I think) for a physicist to synthesize any chemical substance that the chemist writes down. Give the orders and the physicist synthesizes it. How? Put the atoms down where the

◆

The Pleasure of Finding Things Out

chemist says, and so you make the substance. The problems of chemistry and biology can be greatly helped if our ability to see what we are doing, and to do things on an atomic level, is ultimately developed—a development which I think cannot be avoided. Now, you might say, "Who should do this and why should they do it?" Well, I pointed out a few of the economic applications, but I know that the reason that you would do it might be just for fun. But have some fun! Let's have a competition between laboratories. Let one laboratory make a tiny motor which it sends to another lab which sends it back with a thing that fits inside the shaft of the first motor.

High School Competition

Just for the fun of it, and in order to get kids interested in this field, I would propose that someone who has some contact with the high schools think of making some kind of high school competition. After all, we haven't even started in this field, and even the kids can write smaller than has ever been written before. They could have competition in high schools. The Los Angeles high school could send a pin to the Venice high school on which it says, "How's this?" They get the pin back, and in the dot of the "i" it says, "Not so hot."

Perhaps this doesn't excite you to do it, and only economics will do so. Then I want to do something; but I can't do it at the present moment, because I haven't prepared the ground. It is my intention to offer a prize of $1,000 to the first guy who can take the information on the page of a book and put it on an area 1/25,000 smaller in linear scale in such manner that it can be read by an electron microscope.

And I want to offer another prize—if I can figure out how to phrase it so that I don't get into a mess of arguments about definitions—of another $1,000 to the first guy who makes an operating electric motor—a rotating electric motor which can be controlled from the outside and, not counting the lead-in wires, is only 1/64 inch cube.

I do not expect that such prizes will have to wait very long for claimants.

Ultimately Feynman had to make good on both challenges. The following is from the overview to Feynman and Computation, *edited by Anthony J.G. Hey (Perseus, Reading, MA, 1998), reprinted with permission.* Ed.

He paid out on both—the first, less than a year later, to Bill McLellan, a Caltech alumnus, for a miniature motor which satisfied the specifications but which was somewhat of a disappointment to Feynman in that it required no new technical advances. Feynman gave an updated version of his talk in 1983 to the Jet Propulsion Laboratory. He predicted 'that with today's technology we can easily . . . construct motors a fortieth of that size in each dimension, 64,000 times smaller than . . . McLellan's motor, and we can make thousands of them at a time.'

It was not for another 26 years that he had to pay out on the second prize, this time to a Stanford graduate student named Tom Newman. The scale of Feynman's challenge was equivalent to writing all twenty-four volumes of the *Encyclopaedia Brittanica* on the head of a pin: Newman calculated that each individual letter would be only about fifty atoms wide. Using electron-beam lithography when his thesis advisor was out of town, he was eventually able to write the first page of Charles Dickens' *A Tale of Two Cities* at 1/25,000 reduction in scale. Feynman's paper is often credited with starting the field of nanotechnology and there are now regular 'Feynman Nanotechnology Prize' competitions.

6

The Value of Science

Of all its many values, the greatest
must be the freedom to doubt.

✦

In Hawaii, Feynman learns a lesson in humility while touring a
Buddhist temple: "To every man is given the key to the gates of
heaven; the same key opens the gates of hell." This is one of Feynman's
most eloquent pieces, reflecting on science's relevance to the human ex-
perience and vice versa. He also gives a lesson to fellow scientists on
their responsibility to the future of civilization.

From time to time, people suggest to me that scientists
ought to give more consideration to social problems—especially that they should be more responsible in considering the
impact of science upon society. This same suggestion must be
made to many other scientists, and it seems to be generally
believed that if the scientists would only look at these very
difficult social problems and not spend so much time fooling
with the less vital scientific ones, great success would come of
it.

It seems to me that we do think about these problems from
time to time, but we don't put full-time effort into them—the

reason being that we know we don't have any magic formula for solving problems, that social problems are very much harder than scientific ones, and that we usually don't get anywhere when we do think about them.

I believe that a scientist looking at nonscientific problems is just as dumb as the next guy—and when he talks about a nonscientific matter, he will sound as naive as anyone untrained in the matter. Since the question of the value of science is not a scientific subject, this discussion is dedicated to proving my point—by example.

The first way in which science is of value is familiar to everyone. It is that scientific knowledge enables us to do all kinds of things and to make all kinds of things. Of course if we make good things, it is not only to the credit of science; it is also to the credit of the moral choice which led us to good work. Scientific knowledge is an enabling power to do either good or bad—but it does not carry instructions on how to use it. Such power has evident value—even though the power may be negated by what one does.

I learned a way of expressing this common human problem on a trip to Honolulu. In a Buddhist temple there, the man in charge explained a little bit about the Buddhist religion for tourists, and then ended his talk by telling them he had something to say to them that they would *never* forget—and I have never forgotten it. It was a proverb of the Buddhist religion:

"To every man is given the key to the gates of heaven; the same key opens the gates of hell."

What, then, is the value of the key to heaven? It is true that if we lack clear instructions that determine which is the gate to heaven and which the gate to hell, the key may be a dangerous object to use, but it obviously has value. How can we enter heaven without it?

The instructions, also, would be of no value without the key. So it is evident that, in spite of the fact that science could produce enormous horror in the world, it is of value because it *can* produce *something*.

Another value of science is the fun called intellectual enjoyment which some people get from reading and learning and thinking about it, and which others get from working in it. This is a very real and important point and one which is not considered enough by those who tell us it is our social responsibility to reflect on the impact of science on society.

Is this mere personal enjoyment of value to society as a whole? No! But it is also a responsibility to consider the value of society itself. Is it, in the last analysis, to arrange things so that people can enjoy things? If so, the enjoyment of science is as important as anything else.

But I would like *not* to underestimate the value of the worldview which is the result of scientific effort. We have been led to imagine all sorts of things infinitely more marvelous than the imaginings of poets and dreamers of the past. It shows that the imagination of nature is far, far greater than the imagination of man. For instance, how much more remarkable it is for us all to be stuck—half of us upside down—by a mysterious attraction, to a spinning ball that has been swinging in space for billions of years, than to be carried on the back of an elephant supported on a tortoise swimming in a bottomless sea.

I have thought about these things so many times alone that I hope you will excuse me if I remind you of some thoughts that I am sure you have all had—or this type of thought—which no one could ever have had in the past, because people then didn't have the information we have about the world today.

For instance, I stand at the seashore, alone, and start to think. There are the rushing waves . . . mountains of molecules, each stupidly minding its own business . . . trillions apart . . . yet forming white surf in unison.

Ages on ages . . . before any eyes could see . . . year after year . . . thunderously pounding the shore as now. For whom, for what? . . . on a dead planet, with no life to entertain.

Never at rest . . . tortured by energy . . . wasted prodigiously by the sun . . . poured into space. A mite makes the sea roar.

Deep in the sea, all molecules repeat the patterns of one another till complex new ones are formed. They make others like themselves . . . and a new dance starts.

Growing in size and complexity . . . living things, masses of atoms, DNA, protein . . . dancing a pattern ever more intricate.

Out of the cradle onto the dry land . . . here it is standing . . . atoms with consciousness . . . matter with curiosity.

Stands at the sea . . . wonders at wondering . . . I . . . a universe of atoms . . . an atom in the universe.

The Grand Adventure

The same thrill, the same awe and mystery, come again and again when we look at any problem deeply enough. With more knowledge comes deeper, more wonderful mystery, luring one on to penetrate deeper still. Never concerned that the answer may prove disappointing, but with pleasure and confidence we turn over each new stone to find unimagined strangeness leading on to more wonderful questions and mysteries—certainly a grand adventure!

It is true that few unscientific people have this particular type of religious experience. Our poets do not write about it; our artists do not try to portray this remarkable thing. I don't know why. Is nobody inspired by our present picture of the universe? The value of science remains unsung by singers, so you are reduced to hearing—not a song or a poem, but an evening lecture about it. This is not yet a scientific age.

Perhaps one of the reasons is that you have to know how to read the music. For instance, the scientific article says, perhaps, something like this: "The radioactive phosphorus content of the cerebrum of the rat decreases to one-half in a period of two weeks." Now, what does that mean?

It means that phosphorus that is in the brain of a rat (and also in mine, and yours) is not the same phosphorus as it was two weeks ago, but that all of the atoms that are in the brain are being replaced, and the ones that were there before have gone away.

So what is this mind, what are these atoms with consciousness? Last week's potatoes! That is what now can *remember* what was going on in my mind a year ago—a mind which has long ago been replaced.

That is what it means when one discovers how long it takes for the atoms of the brain to be replaced by other atoms, to note that the thing which I call my individuality is only a pattern or dance. The atoms come into my brain, dance a dance, then go out; always new atoms but always doing the same dance, remembering what the dance was yesterday.

The Remarkable Idea

When we read about this in the newspaper, it says, "The scientist says that this discovery may have importance in the cure of cancer." The paper is only interested in the use of the idea, not the idea itself. Hardly anyone can understand the importance of an idea, it is so remarkable. Except that, possibly, some children catch on. And when a child catches on to an idea like that, we have a scientist. These ideas do filter down (in spite of all the conversation about TV replacing thinking), and lots of kids get the spirit—and when they have the spirit you have a scientist. It's too late for them to get the spirit when they are in our universities, so we must attempt to explain these ideas to children.

I would now like to turn to a third value that science has. It is a little more indirect, but not much. The scientist has a lot of experience with ignorance and doubt and uncertainty, and this experience is of very great importance, I think. When a scientist doesn't know the answer to a problem, he is ignorant. When he has a hunch as to what the result is, he is uncertain. And when he is pretty darn sure of what the result is going to be, he is in some doubt. We have found it of paramount importance that in order to progress we must recognize the ignorance and leave room for doubt. Scientific knowledge is a body of statements of varying degrees of certainty—some most unsure, some nearly sure, none *absolutely* certain.

Now, we scientists are used to this, and we take it for granted that it is perfectly consistent to be unsure—that it is possible to live and *not* know. But I don't know whether everyone realizes that this is true. Our freedom to doubt was born of a struggle against authority in the early days of science. It was a very deep and strong struggle. Permit us to question—to doubt, that's all—not to be sure. And I think it is important that we do not forget the importance of this struggle and thus perhaps lose what we have gained. Here lies a responsibility to society.

We are all sad when we think of the wondrous potentialities human beings seem to have, as contrasted with their small accomplishments. Again and again people have thought that we could do much better. They of the past saw in the nightmare of their times a dream for the future. We, of their future, see that their dreams, in certain ways surpassed, have in many ways remained dreams. The hopes for the future today are, in good share, those of yesterday.

Education, for Good and Evil

Once some thought that the possibilities people had were not developed because most of those people were ignorant. With education universal, could all men be Voltaires? Bad can be taught at least as efficiently as good. Education is a strong force, but for either good or evil.

Communications between nations must promote understanding: So went another dream. But the machines of communication can be channeled or choked. What is communicated can be truth or lie. Communication is a strong force also, but for either good or bad.

The applied sciences should free men of material problems at least. Medicine controls diseases. And the record here seems all to the good. Yet there are men patiently working to create great plagues and poisons. They are to be used in warfare tomorrow.

Nearly everybody dislikes war. Our dream today is peace. In peace, man can develop best the enormous possibilities he seems to have. But maybe future men will find that peace, too, can be good and bad. Perhaps peaceful men will drink out of boredom. Then perhaps drink will become the great problem which seems to keep man from getting all he thinks he should out of his abilities.

Clearly, peace is a great force, as is sobriety, as are material power, communication, education, honesty, and the ideals of many dreamers.

We have more of these forces to control than did the ancients. And maybe we are doing a little better than most of them could do. But what we ought to be able to do seems gigantic compared with our confused accomplishments.

Why is this? Why can't we conquer ourselves?

Because we find that even great forces and abilities do not seem to carry with them clear instructions on how to use them. As an example, the great accumulation of understanding as to how the physical world behaves only convinces one that this behavior seems to have a kind of meaninglessness. The sciences do not directly teach good and bad.

Through all ages men have tried to fathom the meaning of life. They have realized that if some direction or meaning could be given to our actions, great human forces would be unleashed. So, very many answers must have been given to the question of the meaning of it all. But they have been of all different sorts, and the proponents of one answer have looked with horror at the actions of the believers in another. Horror, because from a disagreeing point of view all the great potentialities of the race were being channeled into a false and confining blind alley. In fact, it is from the history of the enormous monstrosities created by false belief that philosophers have realized the apparently infinite and wondrous capacities of human beings. The dream is to find the open channel.

What, then, is the meaning of it all? What can we say to dispel the mystery of existence?

If we take everything into account, not only what the ancients knew, but all of what we know today that they didn't know, then I think that we must frankly admit that *we do not know.*

But, in admitting this, we have probably found the open channel.

This is not a new idea; this is the idea of the age of reason. This is the philosophy that guided the men who made the democracy that we live under. The idea that no one really knew how to run a government led to the idea that we should arrange a system by which new ideas could be developed, tried out, tossed out, more new ideas brought in; a trial and error system. This method was a result of the fact that science was already showing itself to be a successful venture at the end of the 18th century. Even then it was clear to socially minded people that the openness of the possibilities was an opportunity, and that doubt and discussion were essential to progress into the unknown. If we want to solve a problem that we have never solved before, we must leave the door to the unknown ajar.

Our Responsibility as Scientists

We are at the very beginning of time for the human race. It is not unreasonable that we grapple with problems. There are tens of thousands of years in the future. Our responsibility is to do what we can, learn what we can, improve the solutions and pass them on. It is our responsibility to leave the men of the future a free hand. In the impetuous youth of humanity, we can make grave errors that can stunt our growth for a long time. This we will do if we say we have the answers now, so young and ignorant; if we suppress all discussion, all criticism, saying, "This is it, boys, man is saved!" and thus doom man for a long time to the chains of authority, confined to the limits of our present imagination. It has been done so many times before.

It is our responsibility as scientists, knowing the great progress and great value of a satisfactory philosophy of ignorance, the great progress that is the fruit of freedom of thought, to proclaim the value of this freedom, to teach how doubt is not to be feared but welcomed and discussed, and to demand this freedom as our duty to all coming generations.

7

Richard P. Feynman's Minority Report to the Space Shuttle *Challenger* Inquiry

✦

When the Space Shuttle Challenger *exploded shortly after its launch on January 28, 1986, six professional astronauts and one school-teacher were tragically killed. The nation was devastated, and NASA was shaken out of its complacency, brought on by years of successful—or at least nonlethal—space missions. A commission was formed, led by Secretary of State William P. Rogers and composed of politicians, astronauts, military men, and one scientist, to investigate the cause of the accident and to recommend steps to prevent such a disaster from ever happening again. The fact that Richard Feynman was that one scientist may have made the difference between answering the question of why the* Challenger *failed and eternal mystery. Feynman was gutsier than most men, not afraid to jet all over the country to talk to the men on the ground, the engineers who had recognized the fact that propaganda was taking the lead over care and safety in the shuttle program. His report, which was perceived by the Commission as embarrassing to NASA, was almost suppressed by*

The Pleasure of Finding Things Out

the Commission, but Feynman fought to have it included; it was relegated to an appendix. When the Commission held a live press conference to answer questions, Feynman did his now-famous tabletop experiment with one of the shuttle's gaskets, or O-rings, and a cup of ice water. It dramatically proved that those key gaskets had failed because the warning of the engineers that it was too cold outside to go ahead with the launch went unheeded by managers eager to impress their bosses with the punctuality of their mission schedule. Here is that historic report.

Introduction

It appears that there are enormous differences of opinion as to the probability of a failure with loss of vehicle and of human life. The estimates range from roughly 1 in 100 to 1 in 100,000. The higher figures come from working engineers, and the very low figures from management. What are the causes and consequences of this lack of agreement? Since 1 part in 100,000 would imply that one could put a Shuttle up each day for 300 years expecting to lose only one, we could more properly ask "What is the cause of management's fantastic faith in the machinery?"

We have also found that certification criteria used in Flight Readiness Reviews often develop a gradually decreasing strictness. The argument that the same risk was flown before without failure is often accepted as an argument for the safety of accepting it again. Because of this, obvious weaknesses are accepted again and again, sometimes without a sufficiently serious attempt to remedy them, or to delay a flight because of their continued presence.

There are several sources of information. There are published criteria for certification, including a history of modifications in the form of waivers and deviations. In addition, the records of the Flight Readiness Reviews for each flight document the arguments used to accept the risks of the flight. Information was obtained from the direct testimony and the reports of the range safety officer, Louis J. Ullian, with respect to the history of success of solid fuel rockets. There was a further study by him (as chairman of the launch abort safety panel [LASP]) in an attempt to determine the risks involved in possible accidents leading to radioactive contamination from attempting to fly a plutonium power supply (RTG) for future planetary missions. The NASA study of the same question is also available. For the history of the Space Shuttle Main Engines, interviews with management and engineers at Marshall, and informal interviews with engineers at Rocketdyne, were made. An independent (Caltech) mechanical engineer who consulted for NASA about engines was also interviewed informally. A visit to Johnson was made to gather information on the reliability of the avionics (computers, sensors, and effectors). Finally there is a report, "A Review of Certification Practices Potentially Applicable to Man-rated Reusable Rocket Engines," prepared at the Jet Propulsion Laboratory by N. Moore, et al., in February 1986, for NASA Headquarters, Office of Space Flight. It deals with the methods used by the FAA and the military to certify their gas turbine and rocket engines. These authors were also interviewed informally.

Solid Fuel Rockets (SRB)

An estimate of the reliability of solid fuel rockets was made by the range safety officer, by studying the experience of all previous rocket flights. Out of a total of nearly 2,900 flights, 121 failed (1 in 25). This includes, however, what may be called early errors, rockets flown for the first few times in which design errors are discovered and fixed. A more reasonable figure for the mature rockets might be 1 in 50. With special care in the selection of the parts and in inspection, a figure of below 1 in 100 might be achieved but 1 in 1,000 is probably not attainable with today's technology. (Since there are two rockets on the Shuttle, these rocket failure rates must be doubled to get Shuttle failure rates from Solid Rocket Booster failure.)

NASA officials argue that the figure is much lower. They point out that these figures are for unmanned rockets but since the Shuttle is a manned vehicle "the probability of mission success is necessarily very close to 1.0." It is not very clear what this phrase means. Does it mean it is close to 1 or that it ought to be close to 1? They go on to explain "Historically this extremely high degree of mission success has given rise to a difference in philosophy between manned space flight programs and unmanned programs; i.e., numerical probability usage versus engineering judgement." (These quotations are from "Space Shuttle Data for Planetary Mission RTG Safety Analysis," pages 3-1, 3-2, February 15, 1985, NASA, JSC.) It is true that if the probability of failure was as low as 1 in 100,000 it would take an inordinate number of tests to determine it (for you would get nothing but a string of perfect flights from which no precise figure, other than that the probability is likely less than the number of such flights in the string so far). But, if the real probability is not so small, flights would show troubles, near failures, and possibly actual failure with a reasonable estimate. In fact, previous NASA experience had shown, on occasion, just such difficulties, near accidents, and accidents, all giving warning that the probability of flight failure was not so very small. The inconsistency of the argument not to determine reliability through historical experience, as the range safety officer did, is that NASA also appeals to history, beginning "Historically this high degree of mission success . . ." Finally, if we are to replace standard numerical probability usage with engineering judgment, why do we find such an enormous disparity between the management estimate and the judgment of the engineers? It would appear that, for whatever purpose, be it for internal or external consumption, the management of NASA exaggerates the reliability of its product, to the point of fantasy.

The history of the certification and Flight Readiness Reviews will not be repeated here. (See other part of Commission reports.) The phenomenon of accepting for flight seals that had shown erosion and blow-by in previous flights is very clear. The *Challenger* flight is an excellent example. There are several references to flights that had gone before. The acceptance and success of these flights is taken as evidence of safety. But erosion and blow-by are not what the design expected. They are warnings that something is wrong. The equipment is not operating as expected, and therefore there is a danger that it can operate with even wider deviations in this unexpected and not thoroughly understood way. The fact that this danger did not lead to a catastrophe before is no guarantee that it will not the next time, unless it is completely understood. When playing Russian roulette the fact that the first shot got off safely is little comfort for the next. The origin and consequences of the erosion and blow-by were not understood. They did not occur equally on all flights and all joints; sometimes more, and sometimes less. Why not sometime, when whatever conditions determined it were right, still more, leading to catastrophe?

In spite of these variations from case to case, officials behaved as if they understood it, giving apparently logical arguments to each other often depending on the "success" of previous flights. For example, in determining if flight 51-L was safe to fly in the face of ring erosion in flight 51-C, it was noted that the erosion depth was only one-third of the radius. It had been noted in an experiment cutting the ring that cutting it as deep as one radius was necessary before the ring failed. Instead of being very concerned that variations of poorly understood conditions might reasonably create a deeper erosion this time, it was asserted, there was "a safety factor of three." This is a strange use of the engineer's term "safety factor." If a bridge is built to withstand a certain load without the beams permanently deforming, cracking, or breaking, it may be designed for the materials used to actually stand up under three times the load. This "safety factor" is to allow for uncertain excesses of load, or unknown extra loads, or weaknesses in the material that might have unexpected flaws, etc. If now the expected load comes on to the new bridge and a crack appears in a beam, this is a failure of the design. There was no safety factor at all; even though the bridge did not actually collapse because the crack only went one-third of the way through the beam. The O-rings of the Solid Rocket Boosters were not designed to erode. Erosion was a clue that something was wrong. Erosion was not something from which safety can be inferred.

There was no way, without full understanding, that one could have confidence that conditions the next time might not produce erosion three times more severe than the time before. Nevertheless, officials fooled themselves into thinking they had such understanding and confidence, in spite of the peculiar variations from case to case. A mathematical model was made to calculate erosion. This was a model based not on physical understanding but on empirical curve fitting. To be more detailed, it was supposed a stream of hot gas impinged on the O-ring material, and the heat was determined at the point of stagnation (so far, with reasonable physical, thermodynamic laws). But to determine how much rubber eroded it was assumed this depended only on this heat by a formula suggested by data on a similar material. A logarithmic plot suggested a straight line, so it was supposed that the erosion varied as the .58 power of the heat, the .58 being determined by a nearest fit. At any rate, adjusting some other numbers, it was determined that the model agreed with the erosion (to a depth of one-third the radius of the ring). There is nothing much so wrong with this as believing the answer! Uncertainties appear everywhere. How strong the gas stream might be was unpredictable, it depended on holes formed in the putty. Blow-by showed that the ring might fail even though not, or only partially, eroded through. The empirical formula was known to be uncertain, for it did not go directly through the very data points by which it was determined. There were a cloud of points some twice above, and some twice below the fitted curve, so erosions twice predicted were reasonable from that cause alone. Similar uncertainties surrounded the other constants in the formula, etc., etc. When using a mathematical model, careful attention must be given to uncertainties in the model.

Liquid Fuel Engine (SSME)

During the flight of 51-L the three Space Shuttle Main Engines all worked perfectly, even, at the last moment, beginning to shut down the engines as the fuel supply began to fail. The question arises, however, as to whether, had it failed, and we were to investigate it in as much detail as we did the

Solid Rocket Booster, we would find a similar lack of attention to faults and a deteriorating reliability. In other words, were the organization weaknesses that contributed to the accident confined to the Solid Rocket Booster sector or were they a more general characteristic of NASA? To that end·the Space Shuttle Main Engines and the avionics were both investigated. No similar study of the Orbiter or the External Tank was made.

The engine is a much more complicated structure than the Solid Rocket Booster, and a great deal more detailed engineering goes into it. Generally, the engineering seems to be of high quality and apparently considerable attention is paid to deficiencies and faults found in operation.

The usual way that such engines are designed (for military or civilian aircraft) may be called the component system, or bottom-up design. First it is necessary to thoroughly understand the properties and limitations of the materials to be used (for turbine blades, for example), and tests are begun in experimental rigs to determine those. With this knowledge larger component parts (such as bearings) are designed and tested individually. As deficiencies and design errors are noted they are corrected and verified with further testing. Since one tests only parts at a time, these tests and modifications are not overly expensive. Finally one works up to the final design of the entire engine, to the necessary specifications. There is a good chance, by this time, that the engine will generally succeed, or that any failures are easily isolated and analyzed because the failure modes, limitations of materials, etc., are so well understood. There is a very good chance that the modifications to the engine to get around the final difficulties are not very hard to make, for most of the serious problems have already been discovered and dealt with in the earlier, less expensive, stages of the process.

Report to the Space Shuttle Challenger *Inquiry*

The Space Shuttle Main Engine was handled in a different manner, top down, we might say. The engine was designed and put together all at once with relatively little detailed preliminary study of the material and components. Then when troubles are found in the bearings, turbine blades, coolant pipes, etc., it is more expensive and difficult to discover the causes and make changes. For example, cracks have been found in the turbine blades of the high pressure oxygen turbopump. Are they caused by flaws in the material, the effect of the oxygen atmosphere on properties of the material, the thermal stresses of startup or shutdown, the vibration and stresses of steady running, or mainly at some resonance at certain speeds, etc.? How long can we run from crack initiation to crack failure, and how does this depend on power level? Using the completed engine as a test bed to resolve such questions is extremely expensive. One does not wish to lose entire engines in order to find out where and how failure occurs. Yet, an accurate knowledge of this information is essential to acquire a confidence in the engine reliability in use. Without detailed understanding, confidence cannot be attained.

A further disadvantage of the top-down method is that, if an understanding of a fault is obtained, a simple fix, such as a new shape for the turbine housing, may be impossible to implement without a redesign of the entire engine.

The Space Shuttle Main Engine is a very remarkable machine. It has a greater ratio of thrust to weight than any previous engine. It is built at the edge of, or outside of, previous engineering experience. Therefore, as expected, many different kinds of flaws and difficulties have turned up. Because, unfortunately, it was built in the top-down manner, they are difficult to find and to fix. The design aim of a lifetime of 55 mission equivalent firings (27,000 seconds of operation, either in a mission of 500 seconds, or on a test stand) has not been obtained. The engine now requires very frequent maintenance and replacement of important parts, such as turbopumps, bearings, sheet metal housings, etc. The high-pressure fuel turbopump had to be replaced every three or four mission equivalents (although that may have been fixed, now) and the high-pressure oxygen turbopump every five or six. This is at most ten percent of the original specification. But our main concern here is the determination of reliability.

In a total of about 250,000 seconds of operation, the engines have failed seriously perhaps 16 times. Engineering pays close attention to these failings and tries to remedy them as quickly as possible. This it does by test studies on special rigs experimentally designed for the flaw in question, by careful inspection of the engine for suggestive clues (like cracks), and by considerable study and analysis. In this way, in spite of the difficulties of top-down design, through hard work many of the problems have apparently been solved.

A list of some of the problems follows. Those followed by an asterisk (*) are probably solved:

> Turbine blade cracks in high pressure fuel turbopumps (HPFTP). (May have been solved.)
> Turbine blade cracks in high pressure oxygen turbopumps (HPOTP).
> Augmented Spark Igniter (ASI) line rupture.*
> Purge check valve failure.*
> ASI chamber erosion.*
> HPFTP turbine sheet metal cracking.
> HPFTP coolant liner failure.*
> Main combustion chamber outlet elbow failure.*
> Main combustion chamber inlet elbow weld offset.*
> HPOTP subsynchronous whirl.*

Flight acceleration safety cutoff system (partial failure in a redundant system).*
Bearing spalling (partially solved).
A vibration at 4,000 Hertz making some engines in-
operable, etc.

Many of these solved problems are the early difficulties of
a new design, for 13 of them occurred in the first 125,000 seconds and only three in the second 125,000 seconds. Natu-
rally, one can never be sure that all the bugs are out, and, for
some, the fix may not have addressed the true cause. Thus, it
is not unreasonable to guess there may be at least one surprise
in the next 250,000 seconds, a probability of 1/500 per engine per mission. On a mission there are three engines, but
some accidents would possibly be contained, and only affect
one engine. The system can abort with only two engines.
Therefore let us say that the unknown surprises do not, even
of themselves, permit us to guess that the probability of mission failure due to the Space Shuttle Main Engine is less than
1/500. To this we must add the chance of failure from known,
but as yet unsolved, problems (those without the asterisk in
the list above). These we discuss below. (Engineers at Rocketdyne, the manufacturer, estimate the total probability as
1/10,000. Engineers at Marshall estimate it as 1/300, while
NASA management, to whom these engineers report, claims
it is 1/100,000. An independent engineer consulting for
NASA thought 1 or 2 per 100 a reasonable estimate.)

The history of the certification principles for these engines
is confusing and difficult to explain. Initially the rule seems
to have been that two sample engines must each have had
twice the time operating without failure, as the operating time
of the engine to be certified (rule of 2x). At least that is the
FAA practice, and NASA seems to have adopted it, originally

expecting the certified time to be 10 missions (hence 20 missions for each sample). Obviously the best engines to use for comparison would be those of greatest total (flight plus test) operating time—the so-called "fleet leaders." But what if a third sample and several others fail in a short time? Surely we will not be safe because two were unusual in lasting longer. The short time might be more representative of the real possibilities, and in the spirit of the safety factor of 2, we should only operate at half the time of the short-lived samples.

The slow shift toward decreasing safety factor can be seen in many examples. We take that of the HPFTP turbine blades. First of all the idea of testing an entire engine was abandoned. Each engine number has had many important parts (like the turbopumps themselves) replaced at frequent intervals, so that the rule must be shifted from engines to components. We accept an HPFTP for a certification time if two samples have each run successfully for twice that time (and of course, as a practical matter, no longer insisting that this time be as large as 10 missions). But what is "successfull"? The FAA calls a turbine blade crack a failure, in order, in practice, to really provide a safety factor greater than 2. There is some time that an engine can run between the time a crack originally starts until the time it has grown large enough to fracture. (The FAA is contemplating new rules that take this extra safety time into account, but only if it is very carefully analyzed through known models within a known range of experience and with materials thoroughly tested. None of these conditions apply to the Space Shuttle Main Engine.)

Cracks were found in many second stage HPFTP turbine blades. In one case three were found after 1,900 seconds, while in another they were not found after 4,200 seconds, although usually these longer runs showed cracks. To follow this story further we shall have to realize that the stress depends a great deal on the power level. The *Challenger* flight was to be at, and previous flights had been at, a power level called 104% of rated power level during most of the time the engines were operating. Judging from some material data it is supposed that at the level 104% of rated power level, the time to crack is about twice that at 109% or full power level (FPL). Future flights were to be at this level because of heavier payloads, and many tests were made at this level. Therefore dividing time at 104% by 2, we obtain units called equivalent full power level (EFPL). (Obviously, some uncertainty is introduced by that, but it has not been studied.) The earliest cracks mentioned above occurred at 1,375 EFPL.

Now the certification rule becomes "limit all second stage blades to a maximum of 1,375 seconds EFPL." If one objects that the safety factor of 2 is lost, it is pointed out that the one turbine ran for 3,800 seconds EFPL without cracks, and half of this is 1,900 so we are being more conservative. We have fooled ourselves in three ways. First we have only one sample, and it is not the fleet leader, for the other two samples of 3,800 or more seconds had 17 cracked blades between them. (There are 59 blades in the engine.) Next we have abandoned the 2x rule and substituted equal time. And finally, 1,375 is where we did see a crack. We can say that no crack had been found below 1,375, but the last time we looked and saw no cracks was 1,100 seconds EFPL. We do not know when the crack formed between these times; for example, cracks may have formed at 1,150 seconds EFPL. (Approximately 2/3 of the blade sets tested in excess of 1,375 seconds EFPL had cracks. Some recent experiments have, indeed, shown cracks as early as 1,150 seconds.) It was important to keep the number high, for the *Challenger* was to fly an engine very close to the limit by the time the flight was over.

The Pleasure of Finding Things Out

Finally it is claimed that the criteria are not abandoned, and the system is safe, by giving up the FAA convention that there should be no cracks, and considering only a completely fractured blade a failure. With this definition no engine has yet failed. The idea is that since there is sufficient time for a crack to grow to fracture we can insure that all is safe by inspecting all blades for cracks. If they are found, replace them, and if none are found we have enough time for a safe mission. This makes the crack problem not a flight safety problem, but merely a maintenance problem.

This may in fact be true. But how well do we know that cracks always grow slowly enough that no fracture can occur in a mission? Three engines have run for long times with a few cracked blades (about 3,000 seconds EFPL) with no blades broken off.

But a fix for this cracking may have been found. By changing the blade shape, shot-peening the surface, and covering with insulation to exclude thermal shock, the blades have not cracked so far.

A very similar story appears in the history of certification of the HPOTP, but we shall not give the details here.

It is evident, in summary, that the Flight Readiness Reviews and certification rules show a deterioration for some of the problems of the Space Shuttle Main Engine that is closely analogous to the deterioration seen in the rules for the Solid Rocket Booster.

Avionics

By "avionics" is meant the computer system on the Orbiter as well as its input sensors and output actuators. At first we will restrict ourselves to the computers proper and not be concerned with the reliability of the input information from the sensors of temperature, pressure, etc., or with whether the computer output is faithfully followed by the actuators of rocket firings, mechanical controls, displays to astronauts, etc.

The computing system is very elaborate, having over 250,000 lines of code. It is responsible, among many other things, for the automatic control of the entire ascent to orbit, and for the descent until well into the atmosphere (below Mach 1) once one button is pushed deciding the landing site desired. It would be possible to make the entire landing automatically (except that the landing gear lowering signal is expressly left out of computer control, and must be provided by the pilot, ostensibly for safety reasons) but such an entirely automatic landing is probably not as safe as a pilot controlled landing. During orbital flight it is used in the control of payloads, in displaying information to the astronauts, and the exchange of information to the ground. It is evident that the safety of flight requires guaranteed accuracy of this elaborate system of computer hardware and software.

In brief, the hardware reliability is ensured by having four essentially independent identical computer systems. Where possible each sensor also has multiple copies, usually four, and each copy feeds all four of the computer lines. If the inputs from the sensors disagree, depending on circumstances, certain averages, or a majority selection is used as the effective input. The algorithm used by each of the four computers is exactly the same, so their inputs (since each sees all copies of the sensors) are the same. Therefore at each step the results in each computer should be identical. From time to time they are compared, but because they might operate at slightly different speeds a system of stopping and waiting at specified times is instituted before each comparison is made. If one of the computers disagrees, or is too late in having its answer ready, the three which do agree are assumed to be correct and the errant computer is taken completely out of the system. If, now, another computer fails, as judged by the agreement of the other two, it is taken out of the system, and the rest of the flight canceled, and descent to the landing site is instituted, controlled by the two remaining computers. It is seen that this is a redundant system since the failure of only one computer does not affect the mission. Finally, as an extra feature of safety, there is a fifth independent computer, whose memory is loaded with only the programs for ascent and descent, and which is capable of controlling the descent if there is a failure of more than two of the computers of the main line of four.

There is not enough room in the memory of the main line computers for all the programs of ascent, descent, and payload programs in flight, so the memory is loaded about four times from tapes, by the astronauts.

Because of the enormous effort required to replace the software for such an elaborate system, and for checking a new system out, no change has been made in the hardware since the system began about fifteen years ago. The actual hardware is obsolete; for example, the memories are of the old ferrite core type. It is becoming more difficult to find manufacturers to supply such old-fashioned computers reliably and of high quality. Modern computers are very much more reliable, can run much faster, simplifying circuits, and allowing more to be done, and would not require so much loading of memory, for their memories are much larger.

The software is checked very carefully in a bottom-up fashion. First, each new line of code is checked, then sections of codes or modules with special function are verified. The scope is increased step by step until the new changes are incorporated into a complete system and checked. This complete output is considered the final product, newly released.

Report to the Space Shuttle Challenger *Inquiry*

But completely independently there is an independent verification group, that takes an adversary attitude to the software development group, and tests and verifies the software as if it were a customer of a delivered product. There is additional verification in using the new programs in simulators, etc. A discovery of an error during the verification testing is considered very serious, and its origin studied very carefully to avoid such mistakes in the future. Such unexpected errors have been found only about six times in all the programming and program changing (for new or altered payloads) that have been done. The principle that is followed is that all the verification is not an aspect of program safety, it is merely a test of that safety, in a non-catastrophic verification. Flight safety is to be judged solely on how well the programs do in the verification tests. A failure here generates considerable concern.

To summarize, then, the computer software checking system and attitude is of highest quality. There appears to be no process of gradually fooling oneself while degrading standards so characteristic of the Solid Rocket Booster or Space Shuttle Main Engine safety systems. To be sure, there have been recent suggestions by management to curtail such elaborate and expensive tests as being unnecessary at this late date in Shuttle history. This must be resisted for it does not appreciate the mutual subtle influences, and sources of error generated by even small changes of one part of a program on another. There are perpetual requests for changes as new payloads and new demands and modifications are suggested by the users. Changes are expensive because they require extensive testing. The proper way to save money is to curtail the number of requested changes, not the quality of testing for each.

One might add that the elaborate system could be very much improved by more modern hardware and programming techniques. Any outside competition would have all the advantages of starting over, and whether that is a good idea for NASA now should be carefully considered.

Finally, returning to the sensors and actuators of the avionics system, we find that the attitude to system failure and reliability is not nearly as good as for the computer system. For example, a difficulty was found with certain temperature sensors sometimes failing. Yet 18 months later the same sensors were still being used, still sometimes failing, until a launch had to be scrubbed because two of them failed at the same time. Even on a succeeding flight this unreliable sensor was used again. Again reaction control systems, the rocket jets used for reorienting and control in flight, still are somewhat unreliable. There is considerable redundancy, but a long history of failures, none of which has yet been extensive enough to seriously affect a flight. The action of the jets is checked by sensors, and if they fail to fire, the computers choose another jet to fire. But they are not designed to fail, and the problem should be solved.

Conclusions

If a reasonable launch schedule is to be maintained, engineering often cannot be done fast enough to keep up with the expectations of originally conservative certification criteria designed to guarantee a very safe vehicle. In these situations, subtly, and often with apparently logical arguments, the criteria are altered so that flights may still be certified in time. They therefore fly in a relatively unsafe condition, with a chance of failure of the order of a percent (it is difficult to be more accurate).

Official management, on the other hand, claims to believe the probability of failure is a thousand times less. One reason for this may be an attempt to assure the government of NASA perfection and success in order to ensure the supply of funds. The other may be that they sincerely believe it to be true, demonstrating an almost incredible lack of communication between themselves and their working engineers.

In any event this has had very unfortunate consequences, the most serious of which is to encourage ordinary citizens to fly in such a dangerous machine, as if it had attained the safety of an ordinary airliner. The astronauts, like test pilots, should know their risks, and we honor them for their courage. Who can doubt that McAuliffe was equally a person of great courage, who was closer to an awareness of the true risk than NASA management would have us believe?

Let us make recommendations to ensure that NASA officials deal in a world of reality in understanding technological weaknesses and imperfections well enough to be actively trying to eliminate them. They must live in reality in comparing the costs and utility of the Shuttle to other methods of entering space. And they must be realistic in making contracts, in estimating costs, and the difficulty of the projects. Only realistic flight schedules should be proposed, schedules that have a reasonable chance of being met. If in this way the government would not support them, then so be it. NASA owes it to the citizens from whom it asks support to be frank, honest, and informative, so that these citizens can make the wisest decisions for the use of their limited resources.

For a successful technology, reality must take precedence over public relations, for nature cannot be fooled.

8

What Is Science?

✦

What is science? It is common sense! Or is it? In April 1966 the master teacher delivered an address to the National Science Teachers' Association in which he gave his fellow teachers lessons on how to teach their students to think like a scientist and how to view the world with curiosity, open-mindedness, and, above all, doubt. This talk is also a tribute to the enormous influence Feynman's father—a uniforms salesman—had on Feynman's way of looking at the world.

I thank Mr. DeRose for the opportunity to join you science teachers. I also am a science teacher. I have too much experience only in teaching graduate students in physics, and as a result of that experience I know that I don't know how to teach.

I am sure that you who are real teachers working at the bottom level of this hierarchy of teachers, instructors of teachers, experts on curricula, also are sure that you, too, don't know how to do it; otherwise you wouldn't bother to come to the Convention.

The subject "What Is Science?" is not my choice. It was Mr. DeRose's subject. But I would like to say that I think that "What Is Science?" is not at all equivalent to "how to teach science," and I must call that to your attention for two reasons. In the first place, from the way that I am preparing to give this lecture, it may seem that I am trying to tell you how to teach science—I am not at all in any way, because I don't know anything about small children. I have one, so I know that I don't know. The other is I think that most of you (because there is so much talk and so many papers and so many experts in the field) have some kind of a feeling of lack of self-confidence. In some way you are always being lectured on how things are not going too well and how you should learn to teach better. I am not going to berate you for the bad works you are doing and indicate how it can definitely be improved; that is not my intention.

As a matter of fact, we have very good students coming into Caltech, and during the years we found them getting better and better. Now how it is done, I don't know. I wonder if you know. I don't want to interfere with the system; it's very good.

Only two days ago we had a conference in which we decided that we don't have to teach a course in elementary quantum mechanics in the graduate school anymore. When I was a student, they didn't even have a course in quantum mechanics in the graduate school it was considered too difficult a subject. When I first started to teach, we had one. Now we teach it to undergraduates. We discover now that we don't have to have elementary quantum mechanics for graduates from other schools. Why is it getting pushed down? Because we are able to teach better in the university, and that is because the students coming up are better trained.

What is science? Of course you all must know, if you teach it. That's common sense. What can I say? If you don't know, every teacher's edition of every textbook gives a complete discussion of the subject. There is some kind of distorted distillation and watered-down and mixed-up words of Francis Bacon from some centuries ago, words which then were supposed to be the deep philosophy of science. But one of the greatest experimental scientists of the time who was really doing something, William Harvey,* said that what Bacon said science was, was the science that a lord chancellor would do. He spoke of making observations, but omitted the vital factor of judgment about what to observe and what to pay attention to.

And so what science is, is not what the philosophers have said it is and certainly not what the teacher editions say it is. What it is, is a problem which I set for myself after I said I would give this talk.

After some time I was reminded of a little poem.

> A centipede was happy quite, until a toad in fun
> Said, "Pray, which leg comes after which?"
> This raised his doubts to such a pitch
> He fell distracted in the ditch
> Not knowing how to run.

All my life, I have been doing science and known what it was, but what I have come to tell you—which foot comes after which—I am unable to do, and furthermore, I am worried by the analogy with the poem, that when I go home I will no longer be able to do any research.

There have been a lot of attempts by the various press reporters to get some kind of a capsule of this talk; I prepared it only a little time ago, so it was impossible; but I can see them all rushing out now to write some sort of headline which says: "The Professor Called the President of NSTA a Toad."

---

*Harvey (1578–1657) discovered the body's circulatory system. *Ed.*

The Pleasure of Finding Things Out

Under these circumstances of the difficulty of the subject, and my dislike of philosophical exposition, I will present it in a very unusual way. I am just going to tell you how I learned what science is. That's a little bit childish. I learned it as a child. I have had it in my blood from the beginning. And I would like to tell you how it got in. This sounds as though I am trying to tell you how to teach, but that is not my intention. I'm going to tell you what science is like by how I learned what science is like.

My father did it to me. When my mother was carrying me, it is reported—I am not directly aware of the conversation—my father said that "if it's a boy, he'll be a scientist." How did he do it? He never told me I should be a scientist. He was not a scientist; he was a businessman, a sales manager of a uniform company, but he read about science and loved it.

When I was very young—the earliest story I know—when I still ate in a high chair, my father would play a game with me after dinner. He had bought a whole lot of old rectangular bathroom floor tiles from someplace in Long Island City. We set them up on end, one next to the other, and I was allowed to push the end one and watch the whole thing go down. So far so good.

Next, the game improved. The tiles were different colors. I must put one white, two blues, one white, two blues, and another white and then two blues—I may want to put another blue, but it must be a white. You recognize already the usual insidious cleverness; first delight him in play, and then slowly inject material of educational value!

Well, my mother, who is a much more feeling woman, began to realize the insidiousness of his efforts and said, "Mel, please let the poor child put a blue tile if he wants to." My father said, "No, I want him to pay attention to patterns. It is the only thing I can do that is mathematics at this earliest level." If I were giving a talk on "what is mathematics?" I would have already answered you. Mathematics is looking for patterns. (The fact is that this education had some effect. We had a direct experimental test at the time I got to kindergarten. We had weaving in those days. They've taken it out; it's too difficult for children. We used to weave colored paper through vertical strips and make patterns. The kindergarten teacher was so amazed that she sent a special letter home to report that this child was very unusual, because he seemed to be able to figure out ahead of time what pattern he was going to get, and made amazingly intricate patterns. So the tile game did do something to me.)

I would like to report other evidence that mathematics is only patterns. When I was at Cornell, I was rather fascinated by the student body, which seems to me was a dilute mixture of some sensible people in a big mass of dumb people studying home economics, etc., including lots of girls. I used to sit in the cafeteria with the students and eat and try to overhear their conversations and see if there was one intelligent word coming out. You can imagine my surprise when I discovered a tremendous thing, it seemed to me.

I listened to a conversation between two girls, and one was explaining that if you want to make a straight line, you see, you go over a certain number to the right for each row you go up, that is, if you go over each time the same amount when you go up a row, you make a straight line. A deep principle of analytic geometry! It went on. I was rather amazed. I didn't realize the female mind was capable of understanding analytic geometry.

She went on and said, "Suppose you have another line coming in from the other side and you want to figure out where they are going to intersect." Suppose on one line you go over two to the right for every one you go up, and the other line goes over three to the right for every one that it goes up, and they start twenty steps apart, etc.—I was flabbergasted. She figured out where the intersection was! It turned out that one girl was explaining to the other how to knit argyle socks.

I, therefore, did learn a lesson: The female mind is capable of understanding analytic geometry. Those people who have for years been insisting (in the face of all obvious evidence to the contrary) that the male and female are equal and capable of rational thought may have something. The difficulty may just be that we have never yet discovered a way to communicate with the female mind. If it is done in the right way, you may be able to get something out of it.

Now I will go on with my own experience as a youngster in mathematics.

Another thing that my father told me—and I can't quite explain it, because it was more an emotion than a telling—was that the ratio of the circumference to the diameter of all circles was always the same, no matter what the size. That didn't seem to me too unobvious, but the ratio had some marvelous property. That was a wonderful number, a deep number, pi.* There was a mystery about this number that I didn't quite understand as a youth, but this was a great thing, and the result was that I looked for π everywhere.

When I was learning later in school how to make the decimals for fractions, and how to make $3\frac{1}{8}$, I wrote 3.125, and thinking I recognized a friend wrote that it equals π, the ratio of circumference to diameter of a circle. The teacher corrected it to 3.1416.

I illustrate these things to show an influence. The idea that there is a mystery, that there is a wonder about the number was important to me, not what the number was.

---

*That is, the Greek lowercase letter π.

Very much later when I was doing experiments in the laboratory—I mean my own home laboratory—fiddling around—no, excuse me, I didn't do experiments, I never did; I just fiddled around. I made radios and gadgets. I fiddled around. Gradually through books and manuals I began to discover there were formulas applicable to electricity in relating the current and resistance, and so on. One day, looking at the formulas in some book or other, I discovered a formula for the frequency of a resonant circuit which was $2 \pi \sqrt{LC}$ where L is the inductance and C the capacitance of the circuit. And there was π, and where was the circle? You laugh, but I was very serious then. π was a thing with circles, and here is π coming out of an electric circuit, where [it stood for] the circle. Do you who laughed know how that π comes about?

I have to love the thing. I have to look for it. I have to think about it. And then I realized, of course, that the coils are made in circles. About a half year later, I found another book which gave the inductance of round coils and square coils, and there were other π's in these formulas. I began to think about it again, and I realized that the π did not come from the circular coils. I understand it better now; but in my heart I still don't quite know where that circle is, where that π comes from. [...]

I would like to say a word or two—may I interrupt my little tale—about words and definitions, because it is necessary to learn the words. It is not science. That doesn't mean just because it is not science that we don't have to teach the words. We are not talking about what to teach; we are talking about what science is. It is not science to know how to change centigrade to Fahrenheit. It's necessary, but it is not exactly science. In the same sense, if you were discussing what art is, you wouldn't say art is the knowledge of the fact that a 3-B pencil is softer than a 2-H pencil. It's a distinct difference. That doesn't mean an art teacher shouldn't teach that, or that an artist gets along very well if he doesn't know that. (Actually you can find out in a minute by trying it; but that's a scientific way that art teachers may not think of explaining.)

In order to talk to each other, we have to have words, and that's all right. It's a good idea to try to see the difference, and it's a good idea to know when we are teaching the tools of science, such as words, and when we are teaching science itself.

To make my point still clearer, I shall pick out a certain science book to criticize unfavorably, which is unfair, because I am sure that with little ingenuity, I can find equally unfavorable things to say about others.

There is a first-grade science book which, in the first lesson of the first grade, begins in an unfortunate manner to teach science, because it starts off on the wrong idea of what science is. There is a picture of a dog, a windable toy dog, and a hand comes to the winder, and then the dog is able to move. Under the last picture, it says "What makes it move?" Later on, there is a picture of a real dog and the question "What makes it move?" Then there is a picture of a motor bike and the question "What makes it move?" and so on.

I thought at first they were getting ready to tell what science was going to be about: physics, biology, chemistry. But that wasn't it. The answer was in the teacher's edition of the book; the answer I was trying to learn is that "energy makes it move."

Now energy is a very subtle concept. It is very, very difficult to get right. What I mean by that is that it is not easy to understand energy well enough to use it right, so that you can deduce something correctly using the energy idea. It is beyond the first grade. It would be equally well to say that "God makes it move," or "spirit makes it move," or "movability makes it move." (In fact equally well to say "energy makes it stop.")

Look at it this way: That's only the definition of energy. It should be reversed. We might say when something can move that it has energy in it, but not "what makes it move is energy." This is a very subtle difference. It's the same with this inertia proposition. Perhaps I can make the difference a little clearer this way:

If you ask a child what makes the toy dog move; if you ask an ordinary human being what makes a toy dog move, that is what you should think about. The answer is that you wound up the spring; it tries to unwind and pushes the gear around. What a good way to begin a science course. Take apart the toy; see how it works. See the cleverness of the gears; see the ratchets. Learn something about the toy, the way the toy is put together, the ingenuity of people, devising the ratchets and other things. That's good. The question is fine. The answer is a little unfortunate, because what they were trying to do is teach a definition of energy. But nothing whatever is learned.

Suppose a student would say, "I don't think energy makes it move." Where does the discussion go from there?

I finally figured out a way to test whether you have taught an idea or you have only taught a definition. Test it this way: You say, "Without using the new word which you have just learned, try to rephrase what you have just learned in your own language." "Without using the word 'energy,' tell me what you know now about the dog's motion." You cannot. So you learned nothing except the definition. You learned nothing about science. That may be all right. You may not want to learn something about science right away. You have to learn definitions. But for the very first lesson is that not possibly destructive?

I think, for lesson number one, to learn a mystic formula for answering questions is very bad. The book has some others—"gravity makes it fall"; "the soles of your shoes wear out because of friction." Shoe leather wears out because it rubs against the sidewalk and the little notches and bumps on the sidewalk grab pieces and pull them off. To simply say it is because of friction is sad, because it's not science.

My father dealt a little bit with energy and used the term after I got a little bit of the idea about it. What he would have done I know, because he did in fact essentially the same thing—though not the same example of the toy dog. He would say, "It moves because the sun is shining," if he wanted to give the same lesson. I would say "No. What has that to do with the sun shining? It moved because I wound up the springs."

"And why, my friend, are you able to move to wind up this spring?"

"I eat."

"What, my friend, do you eat?"

"I eat plants."

"And how do they grow?"

"They grow because the sun is shining."

And it is the same with the dog. What about gasoline? Accumulated energy of the sun which is captured by plants and preserved in the ground. Other examples all end with the sun. And so the same idea about the world that our textbook is driving at is phrased in a very exciting way. All the things that we see that are moving are moving because the sun is shining. It does explain the relationship of one source of energy to another, and it can be denied by the child. He could say, "I don't think it is on account of the sun shining," and you can start a discussion. So there is a difference. (Later I could challenge him with the tides, and what makes the earth turn, and have my hand on mystery again.)

That is just an example of the difference between definitions (which are necessary) and science. The only objection in this particular case was that it was the first lesson. It must certainly come in later, telling you what energy is, but not to such a simple question as "What makes a dog move?" A child should be given a child's answer. "Open it up; let's look at it."

During walks in the woods with my father, I learned a great deal. In the case of birds, for example: Instead of naming them, my father would say, "Look, notice that the bird is always pecking in its feathers. It pecks a lot in its feathers. Why do you think it pecks the feathers?"

I guessed it's because the feathers are ruffled, and he's trying to straighten them out. He said "Okay, when would the feathers get ruffled, or how would they get ruffled?"

"When he flies. When he walks around, it's okay; but when he flies it ruffles the feathers."

Then he would say, "You would guess then when the bird just landed he would have to peck more at his feathers than after he has straightened them out and has been walking around the ground for a while. Okay; let's look."

So we would look, and we would watch, and it turned out, as far as I could make out, that the bird pecked about as much and as often no matter how long he was walking on the ground and not just directly after flight.

So my guess was wrong, and I couldn't guess the right reason. My father revealed the reason.

It is that the birds have lice. There is a little flake that comes off the feather, my father taught me, stuff that can be eaten, and the louse eats it. And then on the louse, there is a little bit of wax in the joints between the sections of the leg that oozes out, and there is a mite that lives in there that can eat that wax. Now the mite has such a good source of food that it doesn't digest it too well, so from the rear end there comes a liquid that has too much sugar, and in that sugar lives a tiny creature, etc.

The facts are not correct. The spirit is correct. First I learned about parasitism, one on the other, on the other, on the other.

Second, he went on to say that in the world whenever there is any source of something that could be eaten to make life go, some form of life finds a way to make use of that source; and that each little bit of leftover stuff is eaten by something.

Now the point of this is that the result of observation, even if I were unable to come to the ultimate conclusion, was a wonderful piece of gold, with a marvelous result. It was something marvelous.

Suppose I were told to observe, to make a list, to write down, to do this, to look, and when I wrote my list down, it was filed with 130 other lists in the back of a notebook. I would learn that the result of observation is relatively dull, that nothing much comes of it.

I think it is very important—at least it was to me—that if you are going to teach people to make observations, you should show that something wonderful can come from them. I learned then what science was about. It was patience. If you looked, and you watched, and you paid attention, you got a great reward from it (although possibly not every time). As a result, when I became a more mature man, I would painstakingly, hour after hour, for years, work on problems—sometimes many years, sometimes shorter times—many of them failing, lots of stuff going into the wastebasket; but every once in a while there was the gold of a new understanding that I had learned to expect when I was a kid, the result of observation. For I did not learn that observation was not worthwhile.

Incidentally, in the forest we learned other things. We would go for walks and see all the regular things, and talk about many things; about the growing plants, the struggle of the trees for light, how they try to get as high as they can, and to solve the problem of getting water higher than 35 or 40 feet, the little plants on the ground that look for the little bits of light that come through, all that growth, and so forth.

One day after we had seen all this, my father took me to the forest again and said, "In all this time we have been looking at the forest, we have only seen half of what is going on, exactly half."

I said, "What do you mean?"

He said, "We have been looking at how all these things grow; but for each bit of growth, there must be the same amount of decay, otherwise the materials would be consumed forever. Dead trees would lie there having used up all the stuff from the air, and the ground, and it wouldn't get back into the ground or the air, and nothing else could grow, because there is no material available. There must be for each bit of growth exactly the same amount of decay."

There then followed many walks in the woods during which we broke up old stumps, saw funny bugs and funguses growing—he couldn't show me bacteria, but we saw the softening effects, and so on. I saw the forest as a process of the constant turning of materials.

There were many such things, description of things, in odd ways. He often started to talk about a thing like this: "Suppose a man from Mars were to come down and look at the world." It's a very good way to look at the world. For example, when I was playing with my electric trains, he told me that there is a great wheel being turned by water which is connected by filaments of copper, which spread out and spread out and spread out in all directions; and then there are little wheels, and all those little wheels turn when the big wheel turns. The relation between them is only that there is copper and iron, nothing else, no moving parts. You turn one wheel here, and all the little wheels all over the place turn, and your train is one of them. It was a wonderful world my father told me about. [...]

What science is, I think, may be something like this: There was on this planet an evolution of life to the stage that there were evolved animals, which are intelligent. I don't mean just human beings, but animals which play and which can learn something from experience (like cats). But at this stage each animal would have to learn from its own experience. They gradually develop, until some animal could learn from experience more rapidly and could even learn from another's experience by watching, or one could show the other, or he saw what the other one did. So there came a possibility that all might learn it, but the transmission was inefficient and they would die, and maybe the one who learned it died, too, before he could pass it on to others.

The question is, is it possible to learn more rapidly what somebody learned from some accident than the rate at which the thing is being forgotten, either because of bad memory or because of the death of the learner or inventors?

So there came a time, perhaps, when for some species the rate at which learning was increased reached such a pitch that suddenly a completely new thing happened; things could be learned by one animal, passed on to another, and another, fast enough that it was not lost to the race. Thus became possible an accumulation of knowledge of the race.

This has been called time-binding. I don't know who first called it this. At any rate, we have here some samples of those animals, sitting here trying to bind one experience to another, each one trying to learn from the other.

What Is Science?

This phenomenon of having a memory for the race, of having an accumulated knowledge passable from one generation to another, was new in the world. But it had a disease in it. It was possible to pass on mistaken ideas. It was possible to pass on ideas which were not profitable for the race. The race has ideas, but they are not necessarily profitable.

So there came a time in which the ideas, although accumulated very slowly, were all accumulations not only of practical and useful things, but great accumulations of all types of prejudices, and strange and odd beliefs.

Then a way of avoiding the disease was discovered. This is to doubt that what is being passed from the past is in fact true, and to try to find out *ab initiio,* again from experience, what the situation is, rather than trusting the experience of the past in the form in which it is passed down. And that is what science is: the result of the discovery that it is worthwhile rechecking by new direct experience, and not necessarily trusting the race experience from the past. I see it that way. That is my best definition.

I would like to remind you all of things that you know very well in order to give you a little enthusiasm. In religion, the moral lessons are taught, but they are not just taught once—you are inspired again and again, and I think it is necessary to inspire again and again, and to remember the value of science for children, for grown-ups, and everybody else, in several ways; not only so that we will become better citizens, more able to control nature and so on. There are other things.

There is the value of the worldview created by science. There is the beauty and the wonder of the world that is discovered through the results of these new experiences. That is to say, the wonders of the content which I just reminded you of; that things move because the sun is shining, which is a deep idea, very strange and wonderful. (Yet, not everything moves because the sun is shining. The earth rotates independent of the sun shining, and the nuclear reactions recently produced energy on the earth, a new source. Probably volcanoes are generally [powered by] a source different from the shining sun.)

The world looks so different after learning science. For example, the trees are made of air, primarily. When they are burned, they go back to air, and in the flaming heat is released the flaming heat of the sun which was bound in to convert the air into trees, and in the ash is the small remnant of the part which did not come from air, that came from the solid earth, instead.

These are beautiful things, and the content of science is wonderfully full of them. They are very inspiring, and they can be used to inspire others.

Another of the qualities of science is that it teaches the value of rational thought, as well as the importance of freedom of thought; the positive results that come from doubting that the lessons are all true. You must here distinguish—especially in teaching—the science from the forms or procedures that are sometimes used in developing science. It is easy to say, "We write, experiment, and observe, and do this or that." You can copy that form exactly. But great religions are dissipated by following form without remembering the direct content of the teaching of the great leaders. In the same way it is possible to follow form and call it science but it is pseudoscience. In this way we all suffer from the kind of tyranny we have today in the many institutions that have come under the influence of pseudoscientific advisers.

We have many studies in teaching, for example, in which people make observations and they make lists and they do statistics, but they do not thereby become established science, established knowledge. They are merely an imitative form of science—like the South Sea Islanders making airfields, radio towers, out of wood, expecting a great airplane to arrive. They even build wooden airplanes of the same shape as they see in the foreigners' airfields around them, but strangely, they don't fly. The result of this pseudoscientific imitation is to produce experts, which many of you are—experts. You teachers who are really teaching children at the bottom of the heap, maybe you can doubt the experts once in a while. Learn from science that you *must* doubt the experts. As a matter of fact, I can also define science another way: Science is the belief in the ignorance of experts.

When someone says science teaches such and such, he is using the word incorrectly. Science doesn't teach it; experience teaches it. If they say to you science has shown such and such, you might ask, "How does science show it—how did the scientists find out—how, what, where?" Not science has shown, but this experiment, this effect, has shown. And you have as much right as anyone else, upon hearing about the experiments (but we must listen to *all* the evidence), to judge whether a reusable conclusion has been arrived at.

In a field which is so complicated that true science is not yet able to get anywhere, we have to rely on a kind of old-fashioned wisdom, a king of definite straightforwardness. I am trying to inspire the teacher at the bottom to have some hope, and some self-confidence in common sense, and natural intelligence. The experts who are leading you may be wrong.

I have probably ruined the system, and the students that are coming into Caltech no longer will be any good. I think we live in an unscientific age in which almost all the buffeting of communications and television words, books, and so on are unscientific. That doesn't mean they are bad, but they are unscientific. As a result, there is a considerable amount of intellectual tyranny in the name of science.

Finally, a man cannot live beyond the grave. Each generation that discovers something from its experience must pass that on, but it must pass that on with a delicate balance of respect and disrespect, so that the race (now that it is aware of the disease to which it is liable) does not inflict its errors too rigidly on its youth, but it does pass on the accumulated wisdom, plus the wisdom that it may not be wisdom.

It is necessary to teach both to accept and to reject the past with a kind of balance that takes considerable skill. Science alone of all the subjects contains within itself the lesson of the danger of belief in the infallibility of the greatest teachers of the preceding generation.

So carry on. Thank you.

9

The Smartest Man in the World

✦

Here is that wonderful 1979 interview of Feynman by Omni maga-
zine. This is Feynman on what he knows and loves best—physics—
and what he loves least, philosophy. ("Philosophers should learn to
laugh at themselves.") Here Feynman discusses the work that earned
him the Nobel Prize, quantum electrodynamics (QED); he then goes
on to cosmology, quarks, and those pesky infinities that gum up so
many equations.

"I think the theory is simply a way to sweep the difficulties
under the rug," Richard Feynman said. "I am, of course, not
sure of that." It sounds like the kind of criticism, ritually tem-
pered, that comes from the audience after a controversial
paper is presented at a scientific conference. But Feynman
was at the podium, delivering a Nobel Prize winner's address.
The theory he was questioning, quantum electrodynamics,
has recently been called "the most precise ever devised"; its
predictions are routinely verified to within one part in a million. When Feynman, Julian Schwinger, and Sin-Itiro Tomon-

aga independently developed it in the 1940s, their colleagues hailed it as "the great cleanup": a resolution of long-standing problems, and a rigorous fusion of the century's two great ideas in physics, relativity and quantum mechanics.

Feynman has combined theoretical brilliance and irreverent skepticism throughout his career. In 1942, after taking his doctorate at Princeton with John Wheeler, he was tapped for the Manhattan Project. At Los Alamos, he was a twenty-five-year-old whiz kid, awed neither by the titans of physics around him (Niels Bohr, Enrico Fermi, Hans Bethe) nor by the top-secret urgency of the project. The security staff was unnerved by his facility at opening safes—sometimes by listening to the tiny movements of the lock mechanism, sometimes by guessing which physical constant the safe's user had chosen as the combination. (Feynman hasn't changed since then; many of his students at Caltech have acquired safe-cracking skills along with their physics.)

After the war, Feynman worked at Cornell University. There, as he recounts in this interview, Bethe was the catalyst for his ideas on resolving "the problem of the infinities." The precise energy levels of electrons in hydrogen atoms, and the forces between the electrons (moving so rapidly that relativistic changes had to be taken into account), had already been the subject of pioneering work for three decades. Every electron, theory asserted, was surrounded by transient "virtual particles" which its mass-energy summoned up from vacuum; those particles in turn summoned up others—and the result was a mathematical cascade which predicted an infinite charge for every electron. Tomonaga had suggested a way around the problem in 1943, and his ideas became known just as Feynman at Cornell and Schwinger at Harvard were making the same crucial step. All three shared the Nobel Prize for Physics in 1965. By then, Feynman's mathematical tools, the "Feynman integrals," and the diagrams he had invented to trace particle interactions were part of the equipment of every theoretical physicist. Mathematician Stanislaw Ulam, another Los Alamos veteran, cites the Feynman diagrams as "a notation that can push thoughts in directions that may prove useful or even novel and decisive." The idea of particles that travel backward in time, for example, is a natural outgrowth of that notation.

In 1950, Feynman moved to Caltech, in Pasadena. His accent is still unmistakably the transplanted New Yorker's, but Southern California seems the appropriate habitat for him: Among the "Feynman stories" his colleagues tell, his fondness for Las Vegas and nightlife in general looms large. "My wife couldn't believe I'd actually accept an invitation to give a speech where I'd have to wear a tuxedo," he says. "I did change my mind a couple of times." In the preface to *The Feynman Lectures on Physics,* widely used as a college text since they were collected and published in 1963, he appears with a maniacal grin, playing a conga drum. (On the bongos, it is said, he can play ten beats with one hand against eleven with the other; try it, and you may decide that quantum electrodynamics is easier.)

Among Feynman's other achievements are his contribution to understanding the phase changes of super-cooled helium, and his work with Caltech colleague Murray Gell-Mann* on the theory of beta decay of atomic nuclei. Both subjects are still far from final resolution, he points out; indeed, he does

---

*(1929–) Winner of the 1969 Nobel Prize in Physics for his contributions and discoveries concerning the classification of elementary particles and their interactions. In 1954 Gell-Mann and G. Zweig introduced the concept of quarks. *Ed.*

not hesitate to call quantum electrodynamics itself a "swindle" that leaves important logical questions unanswered. What kind of man can do work of that caliber while nursing the most penetrating doubts? Read on and find out.

Omni: To someone looking at high-energy physics from the outside, its goal seems to be to find the ultimate constituents of matter. It seems a quest we can trace back to the Greeks' atom, the "indivisible" particle. But with the big accelerators, you get fragments that are more massive than the particles you started with, and maybe quarks that can never be separated. What does that do to the quest?

Feynman: I don't think that ever was the quest. Physicists are trying to find out *how nature behaves;* they may talk carelessly about some "ultimate particle" because that's the way nature looks at a given moment, but . . . Suppose people are exploring a new continent, OK? They see water coming along the ground, they've seen that before, and they call it "rivers." So they say they're exploring to find the headwaters, they go upriver, and sure enough, there they are, it's all going very well. But lo and behold, when they get up far enough they find the whole system's different: There's a great big lake, or springs, or the rivers run in a circle. You might say, "Aha! They've failed!" but not at all! The *real* reason they were doing it was to explore the land. If it turned out not to be headwaters, they might be slightly embarrassed at their carelessness in explaining themselves, but no more than that. As long as it looks like the way things are built is wheels within wheels, then you're looking for the innermost wheel—but it might not be that way, in which case you're looking for whatever the hell it is that you find!

Omni: But surely you must have some guess about what you'll find; there are bound to be ridges and valleys and so on. . . ?

Feynman: Yeah, but what if when you get there it's all clouds? You can expect certain things, you can work out theorems about the topology of watersheds, but what if you find a kind of mist, maybe, with things coagulating out of it, with no way to distinguish the land from the air? The whole idea you started with is gone! That's the kind of exciting thing that happens from time to time. One is presumptuous if one says, "We're going to find the ultimate particle, or the unified field laws," or "*the*" anything. If it turns out surprising, the scientist is even more delighted. You think he's going to say, "Oh, it's not like I expected, there's no ultimate particle, I don't want to explore it"? No, he's going to say, "What the hell *is* it, then?"

Omni: You'd rather see that happen?

Feynman: Rather doesn't make any difference: I get what I get. You can't say it's *always* going to be surprising, either; a few years ago I was very skeptical about the gauge theories,* partly because I expected the strong nuclear interaction to be more different from electrodynamics than it now looks. I was expecting mist, and now it looks like ridges and valleys after all.

Omni: Are physical theories going to keep getting more abstract and mathematical? Could there be today a theorist like Faraday in the early nineteenth century, not mathematically sophisticated but with a very powerful intuition about physics?

Feynman: I'd say the odds are strongly against it. For one thing, you need the math just to understand what's been done so far. Beyond that, the behavior of subnuclear systems is so strange compared to the ones the brain evolved to deal

---

*Theories in particle physics that describe the various interactions between subatomic particles. *Ed.*

with that the analysis *has* to be very abstract: To understand ice, you have to understand things that are themselves very unlike ice. Faraday's models were mechanical—springs and wires and tense bands in space—and his images were from basic geometry. I think we've understood all we can from that point of view; what we've found in this century is different enough, obscure enough, that further progress will require a lot of math.

Omni: Does that limit the number of people who can contribute, or even understand what's being done?

Feynman: Or else somebody will develop a way of thinking about the problems so that we can understand them more easily. Maybe they'll just teach it earlier and earlier. You know, it's not true that what is called "abstruse" math is so difficult. Take something like computer programming, and the careful logic needed for that—the kind of thinking that mama and papa would have said was only for professors. Well, now it's part of a lot of daily activities, it's a way to make a living; their children get interested and get hold of a computer and they're doing the most crazy, wonderful things!

Omni: . . . with ads for programming schools on every matchbook!

Feynman: Right. I don't believe in the idea that there are a few peculiar people capable of understanding math, and the rest of the world is normal. Math is a human discovery, and it's no more complicated than humans can understand. I had a calculus book once that said, "What one fool can do, another can." What we've been able to work out about nature may look abstract and threatening to someone who hasn't studied it, but it was fools who did it, and in the next generation, all the fools will understand it.

There's a tendency to pomposity in all this, to make it all deep and profound. My son is taking a course in philosophy, and last night we were looking at something by Spinoza—and there was the most childish reasoning! There were all these Attributes, and Substances, all this meaningless chewing around, and we started to laugh. Now, how could we do that? Here's this great Dutch philosopher, and we're laughing at him. It's because there was no excuse for it! In that same period there was Newton, there was Harvey studying the circulation of the blood, there were people with methods of analysis by which progress was being made! You can take every one of Spinoza's propositions, and take the contrary propositions, and look at the world—and you can't tell which is right. Sure, people were awed because he had the courage to take on these great questions, but it doesn't do any good to have the courage if you can't get anywhere with the question.

Omni: In your published lectures, the philosopher's comments on science come in for some lumps. . .

Feynman: It isn't the philosophy that gets me, it's the pomposity. If they'd just *laugh* at themselves! If they'd just say, "I think it's like this, but von Leipzig thought it was like that, and he had a good shot at it, too." If they'd explain that this is their best guess. . . But so few of them do; instead, they seize on the possibility that there may not be any ultimate fundamental particle, and say that you should stop work and ponder with great profundity. "You haven't thought deeply enough, first let me define the world for you." Well, I'm going to investigate it *without* defining it!

Omni: How do you know which problem is the right size to attack?

Feynman: When I was in high school, I had this notion that you could take the importance of the problem and multiply by your chance of solving it. You know how a technically minded kid is, he likes the idea of optimizing everything . . . anyway, if you can get the right combination of those factors, you don't spend your life getting nowhere with a profound problem, or solving lots of small problems that others could do just as well.

Omni: Let's take the problem that won the Nobel Prize for you, Schwinger, and Tomonaga. Three different approaches: Was that problem especially ripe for solution?

Feynman: Well, quantum electrodynamics had been invented in the late 1920s by Dirac and others, just after quantum mechanics itself. They had it fundamentally correct, but when you went to calculate answers you ran into complicated equations that were very hard to solve. You could get a good first-order approximation, but when you tried to refine it with corrections these infinite quantities started to crop up. Everybody knew that for twenty years; it was in the back of all the books on quantum theory.

Then we got the results of experiments by Lamb* and Rutherford† on the shifts in energy of the electron in hydrogen atoms. Until then, the rough prediction had been good enough, but now you had a very precise number: 1060 megacycles or whatever. And everybody said dammit, this problem has to be solved . . . they'd known the theory had problems, but now there was this very precise figure.

So Hans Bethe took this figure and made some estimates of how you could avoid the infinities by subtracting this effect from that effect, so the quantities that would tend to go to in-

---

*Willis Lamb (1913–), winner of the 1955 Nobel Prize in Physics for his discoveries concerning the fine structure of the hydrogen spectrum. *Ed.*

†Lord Ernest Rutherford, 1st Baron Rutherford of Nelson (1871–1937), winner of the 1908 Nobel Prize in Chemistry for his investigations into the disintegration of the elements, and the chemistry of radioactive substances. *Ed.*

finity were stopped short, and they'd probably stop in this order of magnitude, and he came out with something around 1000 megacycles. I remember, he'd invited a bunch of people to a party at his house, at Cornell, but he'd been called away to do some consulting. He called up during the party and told me he'd figured this out on the train. When he came back he gave a lecture on it, and showed how this cut-off procedure avoided the infinities, but was still very *ad hoc* and confusing. He said it would be good if someone could show how it could be cleaned up. I went up to him afterwards and said, "Oh, that's easy, I can do that." See, I'd started to get ideas on this when I was a senior at MIT. I'd even cooked up an answer then—wrong, of course. See, this is where Schwinger and Tomonaga and I came in, in developing a way to turn this kind of procedure into solid analysis—technically, to maintain relativistic invariance all the way through. Tomonaga had already suggested how it could be done, and at this same time Schwinger was developing his own way.

So I went to Bethe with my way of doing it. The funny thing was, I didn't know how to do the simplest practical problems in this area—I should have learned long before, but I'd been busy playing with my own theory—so I didn't know how to find out if my ideas worked. We did it together on the blackboard, and it was wrong. Even worse than before. I went home and thought and thought, and decided I had to learn to solve examples. So I did, and I went back to Bethe and we tried it and it worked! We've never been able to figure out what went wrong the first time . . . some dumb mistake.

Omni: How far had it set you back?

Feynman: Not much: maybe a month. It did me good, because I reviewed what I'd done and convinced myself that it had to work, and that these diagrams I'd invented to keep things straight were really OK.

The Pleasure of Finding Things Out

Omni: Did you realize at that time that they'd be called "Feynman diagrams," that they'd be in the books?

Feynman: No, not—I do remember one moment. I was in my pajamas, working on the floor with papers all around me, these funny-looking diagrams of blobs with lines sticking out. I said to myself, wouldn't it be funny if these diagrams really are useful, and other people start using them, and *Physical Review* has to print these silly pictures? Of course, I couldn't foresee—in the first place, I had no idea how many of these pictures there'd be in *Physical Review,* and in the second place, it never occurred to me that with everybody using them, they wouldn't look funny anymore. . .

[At this point the interview adjourned to Professor Feynman's office, where the tape recorder refused to start again. The cord, power switch, "record" button, all were in order; then Feynman suggested taking the tape cassette out and putting it in again.]

Feynman: There. See, you just have to know about the world. Physicists know about the world.

Omni: Take it apart and put it back together?

Feynman: Right. There's always a little dirt, or infinity, or something.

Omni: Let's follow that up. In your lectures, you say that our physical theories do well at uniting various classes of phenomena, and then X-rays or mesons or the like show up; "There are always many threads hanging out in all directions." What are some of the loose threads you see in physics today?

Feynman: Well, there are the masses of the particles: The gauge theories give beautiful patterns for the interactions, but not for the masses, and we need to understand this irregular set of numbers. In the strong nuclear interaction, we have this theory of colored* quarks and gluons, very precise and completely stated, but with very few hard predictions. It's technically very difficult to get a sharp test of the theory, and that's a challenge. I feel passionately that that's a loose thread; while there's no evidence in conflict with the theory, we're not likely to make much progress until we can check hard predictions with hard numbers.

Omni: What about cosmology? Dirac's suggestion that the fundamental constants change with time, or the idea that physical law was different at the instant of the Big Bang?

Feynman: That would open up a lot of questions. So far, physics has tried to find laws and constants without asking where they came from, but we may be approaching the point where we'll be forced to consider history.

Omni: Do you have any guesses on that?

Feynman: No.

Omni: None at all? No leaning either way?

Feynman: No, really. That's the way I am about almost everything. Earlier, you didn't ask whether I thought that there's a fundamental particle, or whether it's all mist; I would have told you that I haven't the slightest idea. Now, in order to work hard on something, you have to get yourself believing that the answer's over *there,* so you'll dig hard there, right? So you temporarily prejudice or predispose yourself— but all the time, in the back of your mind, you're laughing. Forget what you hear about science without prejudice. Here, in an interview, talking about the Big Bang, I have no prejudices—but when I'm working, I have a lot of them.

---

*"Color" is actually a name physicists gave to a certain property of quarks and gluons not because they have any actual color, but for want of a better name for a new property of elementary particles. *Ed.*

Omni: Prejudices in favor of . . . what? Symmetry, simplicity. . . ?

Feynman: In favor of my mood of the day. One day I'll be convinced there's a certain type of symmetry that everybody believes in, the next day I'll try to figure out the consequences if it's not, and everybody's crazy but me. But the thing that's unusual about good scientists is that while they're doing whatever they're doing, they're not so sure of themselves as others usually are. They can live with steady doubt, think "maybe it's so" and act on that, all the time knowing it's only "maybe." Many people find that difficult; they think it means detachment or coldness. It's not coldness! It's a much deeper and warmer understanding, and it means you can be digging somewhere where you're temporarily convinced you'll find the answer, and somebody comes up and says, "Have you seen what they're coming up with over there?", and you look up and say *"Jeez! I'm in the wrong place!"* It happens all the time.

Omni: There's another thing that seems to happen a lot in modern physics: the discovery of applications for kinds of mathematics that were previously "pure," such as matrix algebra or group therapy. Are physicists more receptive now than they used to be? Is the time lag less?

Feynman: There never was any time lag. Take Hamilton's* quaternions: the physicists threw away most of this very powerful mathematical system, and kept only the part—the mathematically almost trivial part—that became vector analysis. But when the whole power of quaternions *was* needed, for quantum mechanics, Pauli† re-invented the system on the

---

*Sir William Rowan Hamilton (1805–1865), Irish mathematician who invented quaternions, an alternate construct to tensor and vector analysis. *Ed.*

†Wolfgang Pauli (1900–1958), winner of the 1945 Nobel Prize in Physics for his discovery of exclusion principle. *Ed.*

spot in a new form. Now, you can look back and say that Pauli's spin matrices and operators were nothing but Hamilton's quaternions ... but even if physicists had kept the system in mind for ninety years, it wouldn't have made more than a few weeks' difference.

Say you've got a disease, Werner's granulomatosis or whatever, and you look it up in a medical reference book. You may well find that you then know more about it than your doctor does, although he spent all that time in medical school ... you see? It's much easier to learn about some special, restricted topic than a whole field. The mathematicians are exploring in all directions, and it's quicker for a physicist to catch up on what he needs than to try to keep up with everything that might conceivably be useful. The problem I was mentioning earlier, the difficulties with the equations in the quark theories—it's the physicists' problem, and we're going to solve it, and maybe when we solve it we'll be doing mathematics. It's a marvelous fact, and one I don't understand, that the mathematicians had investigated groups and so on before they turned up in physics—but in regard to the speed of progress in physics, I don't think it's all that significant.

Omni: One more question from your lectures: you say there that "the next great era of awakening of human intellect may well produce a method of understanding the *qualitative* content of equations." What do you mean by that?

Feynman: In that passage I was talking about the Schrödinger* equation. Now, you can get from that equation to atoms bonding in molecules, chemical valences—but when

---

*Erwin Schrödinger (1887–1961), winner (with P.A.M. Dirac) of the 1933 Nobel Prize in Physics for the discovery of new, productive forms of atomic theory. *Ed.*

you look at the equation, you can see nothing of the wealth of phenomena that the chemists know about; or the idea that quarks are permanently bound so you can't get a free quark—maybe you can and maybe you can't, but the point is that when you look at the equations that supposedly describe quark behavior, you can't see why it should be so. Look at the equations for the atomic and molecular forces in water, and you can't see the way water behaves; you can't see turbulence.

Omni: That leaves the people with questions about turbulence—the meteorologists and oceanographers and geologists and airplane designers—kind of up the creek, doesn't it?

Feynman: Absolutely. And it might be one of those up-the-creek people who'll get so frustrated he'll figure it out, and at that point he'll be doing physics. With turbulence, it's not just a case of physical theory being able to handle only simple cases—we can't do *any*. We have no good fundamental theory at all.

Omni: Maybe it's the way the textbooks are written, but few people outside science appear to know just how quickly real, complicated physical problems get out of hand as far as theory is concerned.

Feynman: That's very bad education. The lesson you learn as you grow older in physics is that what we can do is a very small fraction of what there is. Our theories are really very limited.

Omni: Do physicists vary greatly in their ability to see the qualitative consequences of an equation?

Feynman: Oh, yes—but nobody is very good at it. Dirac said that to *understand* a physical problem means to be able to see the answer without solving equations. Maybe he exaggerated; maybe solving equations is experience you need to gain understanding—but until you do understand, you're just solving equations.

The Smartest Man in the World

Omni: As a teacher, what can you do to encourage that ability?

Feynman: I don't know. I have no way to judge the degree to which I'm getting across to my students.

Omni: Will a historian of science someday trace the careers of your students as others have done with the students of Rutherford and Niels Bohr and Fermi?

Feynman: I doubt it. I'm disappointed with my students all the time. I'm not a teacher who knows what he's doing.

Omni: But you can trace influences the other way, say, the influence on you of Hans Bethe or John Wheeler. . . ?

Feynman: Sure. But I don't know the effect *I'm* having. Maybe it's just my character: I don't know. I'm not a psychologist or sociologist, I don't know how to understand people, including myself. You ask, how can this guy teach, how can he be motivated if he doesn't know what he's doing? As a matter of fact, I love to teach. I like to think of new ways of looking at things as I explain them, to make them clearer—but maybe I'm not making them clearer. Probably what I'm doing is entertaining myself.

I've learned how to live without knowing. I don't have to be sure I'm succeeding, and as I said before about science, I think my life is fuller because I realize that I don't know what I'm doing. I'm delighted with the width of the world!

Omni: As we came back to the office, you stopped to discuss a lecture on color vision you'll be giving. That's pretty far from fundamental physics, isn't it? Wouldn't a physiologist say you were "poaching"?

Feynman: Physiology? It has to be physiology? Look, give me a little time and I'll give a lecture on anything in physiology. I'd be delighted to study it and find out all about it, because I can *guarantee* you it would be very interesting. I don't know anything, but I do know that *everything is interesting* if you go into it deeply enough.

The Pleasure of Finding Things Out

My son is like that, too, although he's much wider in his interests than I was at his age. He's interested in magic, in computer programming, in the history of the early church, in topology—oh, he's going to have a terrible time, there are so many interesting things. We like to sit down and talk about how different things could be from what we expected; take the Viking landers on Mars, for example, we were trying to think how many ways there could be life that they *couldn't* find with that equipment. Yeah, he's a lot like me, so at least I've passed on this idea that everything is interesting to at least one other person.

Of course, I don't know if that's a good thing or not. . . . You see?

10

Cargo Cult Science: Some Remarks on Science, Pseudoscience, and Learning How to Not Fool Yourself

The 1974 Caltech Commencement Address

Question: *What do witch doctors, ESP, South Sea Islanders, rhinoceros horns, and Wesson Oil have to do with college graduation? Answer: They're all examples the crafty Feynman uses to convince departing graduates that honesty in science is more rewarding than all the kudos and temporary successes in the world. In this address to Caltech's class of 1974, Feynman gives a lesson in scientific integrity in the face of peer pressure and glowering funding agencies.*

During the Middle Ages there were all kinds of crazy ideas, such as that a piece of rhinoceros horn would increase potency. (Another crazy idea of the Middle Ages is these hats we have on today–which is too loose in my case.) Then a method was discovered for separating the ideas—which was to try one to see if it worked, and if it didn't work, to eliminate it. This method became organized, of course, into science. And it developed very well, so that we are now in the scientific age. It is such a scientific age, in fact, that we have difficulty in understanding how witch doctors could *ever* have existed, when nothing that they proposed ever really worked—or very little of it did.

But even today I meet lots of people who sooner or later get me into a conversation about UFOs, or astrology, or some form of mysticism, expanded consciousness, new types of awareness, ESP, and so forth. And I've concluded that it's *not* a scientific world.

Most people believe so many wonderful things that I decided to investigate why they did. And what has been referred to as my curiosity for investigation has landed me in a difficulty where I found so much junk to talk about that I can't do it in this talk. I'm overwhelmed. First I started out by investigating various ideas of mysticism, and mystic experiences. I went into isolation tanks (they're dark and quiet and you float in Epsom salts) and got many hours of hallucinations, so I know something about that. Then I went to Esalen, which is a hotbed of this kind of thought (it's a wonderful place; you should go visit there). Then I became overwhelmed. I didn't realize how *much* there was.

I was sitting, for example, in a hot bath and there's another guy and a girl in the bath. He says to the girl, "I'm learning massage and I wonder if I could practice on you?" She says OK, so she gets up on a table and he starts off on her foot—working on her big toe and pushing it around. Then he turns to what is apparently his instructor, and says, "I feel a kind of dent. Is that the pituitary?" And she says, "No, that's not the way it feels." I say, "You're a hell of a long way from the pituitary, man." And they both looked at me—I had blown my cover, you see—and she said, "It's reflexology." So I closed my eyes and appeared to be meditating.

That's just an example of the kind of things that overwhelm me. I also looked into extrasensory perception and PSI phenomena, and the latest craze there was Uri Geller, a man who is supposed to be able to bend keys by rubbing them with his finger. So I went to his hotel room, on his invitation, to see a demonstration of both mind reading and bending keys. He didn't do any mind reading that succeeded; nobody can read my mind, I guess. And my boy held a key and Geller rubbed it, and nothing happened. Then he told us it works better under water, and so you can picture all of us standing in the bathroom with the water turned on and the key under it, and him rubbing the key with his finger. Nothing happened. So I was unable to investigate that phenomenon.

But then I began to think, what else is there that we believe? (And I thought then about the witch doctors, and how easy it would have been to check on them by noticing that nothing really worked.) So I found things that even *more* people believe, such as that we have some knowledge of how to educate. There are big schools of reading methods and mathematics methods, and so forth, but if you notice, you'll see the reading scores keep going down—or hardly going up—in spite of the fact that we continually use these same people to improve the methods. *There's* a witch doctor remedy that doesn't work. It ought to be looked into; how do they know that their method should work? Another example is how to treat criminals. We obviously have made no progress—lots of theory, but no progress—in decreasing the amount of crime by the method that we use to handle criminals.

Yet these things are said to be scientific. We study them. And I think ordinary people with commonsense ideas are intimidated by this pseudoscience. A teacher who has some good idea of how to teach her children to read is forced by the school system to do it some other way—or is even fooled by the school system into thinking that her method is not necessarily a good one. Or a parent of bad boys, after disciplining them in one way or another, feels guilty for the rest of her life because she didn't do "the right thing," according to the experts.

So we really ought to look into theories that don't work, and science that isn't science.

I tried to find a principle for discovering more of these kinds of things, and came up with the following system. Anytime you find yourself in a conversation at a cocktail party in which you do not feel uncomfortable that the hostess might come around and say, "Why are you fellows talking shop?" or that your wife will come around and say, "Why are you flirting again?"—then you can be sure you are talking about something about which nobody knows anything.

Using this method, I discovered a few more topics that I had forgotten—among them the efficacy of various forms of psychotherapy. So I began to investigate through the library, and so on, and I have so much to tell you that I can't do it all. I will have to limit myself to just a few little things. I'll concentrate on the things more people believe in. Maybe I will give a series of speeches next year on all these subjects. It will take a long time.

I think the educational and psychological studies I mentioned are examples of what I would like to call Cargo Cult Science. In the South Seas there is a Cargo Cult of people. During the war they saw airplanes land with lots of good materials, and they want the same thing to happen now. So they've arranged to make things like runways, to put fires along the sides of the runways, to make a wooden hut for a man to sit in, with two wooden pieces on his head like headphones and bars of bamboo sticking out like antennas—he's the controller—and they wait for the airplanes to land. They're doing everything right. The form is perfect. It looks exactly the way it looked before. But it doesn't work. No airplanes land. So I call these things Cargo Cult Science, because they follow all the apparent precepts and forms of scientific investigation, but they're missing something essential, because the planes don't land.

Now it behooves me, of course, to tell you what they're missing. But it would be just about as difficult to explain to the South Sea Islanders how they have to arrange things so that they get some wealth in their system. It is not something simple like telling them how to improve the shapes of the earphones. But there is *one* feature I notice that is generally missing in Cargo Cult Science. That is the idea that we all hope you have learned in studying science in school—we never explicitly say what this *is,* but just hope that you catch on by all the examples of scientific investigation. It is interesting, therefore, to bring it out now and speak of it explicitly. It's a kind of scientific integrity, a principle of scientific thought that corresponds to a kind of utter honesty—a kind of leaning over backwards. For example, if you're doing an experiment, you should report everything that you think might make it invalid—not only what you think is right about it: other causes that could possibly explain your results; and things you thought of that you've eliminated by some other experiment, and how they worked—to make sure the other fellow can tell they have been eliminated.

Details that could throw doubt on your interpretation must be given, if you know them. You must do the best you can—if you know anything at all wrong, or possibly wrong—to explain it. If you make a theory, for example, and advertise it, or put it out, then you must also put down all the facts that disagree with it, as well as those that agree with it. There is also a more subtle problem. When you have put a lot of ideas together to make an elaborate theory, you want to make sure, when explaining what it fits, that those things it fits are not just the things that gave you the idea for the theory; but that the finished theory makes something else come out right, in addition.

In summary, the idea is to try to give *all* of the information to help others to judge the value of your contribution; not just the information that leads to judgment in one particular direction or another.

The easiest way to explain this idea is to contrast it, for example, with advertising. Last night I heard that Wesson Oil doesn't soak through food. Well, that's true. It's not dishonest; but the thing I'm talking about is not just a matter of not being dishonest, it's a matter of scientific integrity, which is another level. The fact that should be added to that advertising statement is that *no* oils soak through food, if operated at a certain temperature. If operated at another temperature, they *all* will—including Wesson Oil. So it's the implication which has been conveyed, not the fact, which is true, and the difference is what we have to deal with.

We've learned from experience that the truth will out. Other experimenters will repeat your experiment and find out whether you were wrong or right. Nature's phenomena will agree or they'll disagree with your theory. And, although you may gain some temporary fame and excitement, you will not gain a good reputation as a scientist if you haven't tried to be very careful in this kind of work. And it's this type of integrity, this kind of care not to fool yourself, that is missing to a large extent in much of the research in Cargo Cult Science.

Cargo Cult Science: The 1974 Caltech Commencement Address

A great deal of their difficulty is, of course, the difficulty of the subject and the inapplicability of the scientific method to the subject. Nevertheless, it should be remarked that this is not the only difficulty. That's *why* the planes don't land—but they don't land.

We have learned a lot from experience about how to handle some of the ways we fool ourselves. One example: Millikan measured the charge on an electron by an experiment with falling oil drops and got an answer which we now know not to be quite right. It's a little bit off, because he had the incorrect value for the viscosity of air. It's interesting to look at the history of measurements of the charge of the electron, after Millikan. If you plot them as a function of time, you find that one is a little bigger than Millikan's, and the next one's a little bit bigger than that, and the next one's a little bit bigger than that, until finally they settle down to a number which is higher.

Why didn't they discover that the new number was higher right away? It's a thing that scientists are ashamed of—this history—because it's apparent that people did things like this: When they got a number that was too high above Millikan's, they thought something must be wrong—and they would look for and find a reason why something might be wrong. When they got a number closer to Millikan's value, they didn't look so hard. And so they eliminated the numbers that were too far off, and did other things like that. We've learned those tricks nowadays, and now we don't have that kind of a disease.

But this long history of learning how to not fool ourselves—of having utter scientific integrity—is, I'm sorry to say, something that we haven't specifically included in any particular course that I know of. We just hope you've caught on by osmosis.

The first principle is that you must not fool yourself—and you are the easiest person to fool. So you have to be very careful about that. After you've not fooled yourself, it's easy not to fool other scientists. You just have to be honest in a conventional way after that.

I would like to add something that's not essential to the scientist, but something I kind of believe, which is that you should not fool the layman when you're talking as a scientist. I am not trying to tell you what to do about cheating on your wife, or fooling your girlfriend, or something like that, when you're not trying to be a scientist, but just trying to be an ordinary human being. We'll leave those problems up to you and your rabbi. I'm talking about a specific, extra type of integrity that is not lying, but bending over backwards to show how you're maybe wrong, that you ought to do when acting as a scientist. And this is our responsibility as scientists, certainly to other scientists, and I think to laymen.

For example, I was a little surprised when I was talking to a friend who was going to go on the radio. He does work on cosmology and astronomy, and he wondered how he would explain what the applications of this work were. "Well," I said, "there aren't any." He said, "Yes, but then we won't get support for more research of this kind." *I* think that's kind of dishonest. If you're representing yourself as a scientist, then you should explain to the layman what you're doing—and if they don't want to support you under these circumstances, then that's their decision.

One example of the principle is this: If you've made up your mind to test a theory, or you want to explain some idea, you should always decide to publish it whichever way it comes out. If we only publish results of a certain kind, we can make the argument look good. We must publish *both* kinds of results. For example—let's take advertising again—suppose

◆

Cargo Cult Science: The 1974 Caltech Commencement Address

some particular cigarette has some particular property, like low nicotine. It's published widely by the company that this means it is good for you—they don't say, for instance, that the tars are a different proportion, or that something else is the matter with the cigarette. In other words, publication probability depends upon the answer. That should not be done.

I say that's also important in giving certain types of government advice. Supposing a senator asked you for advice about whether drilling a hole should be done in his state; and you decide it would be better in some other state. If you don't publish such a result, it seems to me you're not giving scientific advice. You're being used. If your answer happens to come out in the direction the government or the politicians like, they can use it as an argument in their favor; if it comes out the other way, they don't publish it at all. That's not giving scientific advice.

Other kinds of errors are more characteristic of poor science. When I was at Cornell, I often talked to the people in the psychology department. One of the students told me she wanted to do an experiment that went something like this—I don't remember it in detail, but it had been found by others that under certain circumstances, X, rats did something, A. She was curious as to whether, if she changed the circumstances to Y, they would still do A. So her proposal was to do the experiment under circumstances Y and see if they still did A.

I explained to her that it was necessary first to repeat in her laboratory the experiment of the other person—to do it under condition X to see if she could also get result A—and then change to Y and see if A changed. Then she would know that the real difference was the thing she thought she had under control.

She was very delighted with this new idea, and went to her professor. And his reply was, no, you cannot do that, because the experiment has already been done and you would be wasting time. This was in about 1935 or so, and it seems to have been the general policy then to not try to repeat psychological experiments, but only to change the conditions and see what happens.

Nowadays there's a certain danger of the same thing happening, even in the famous field of physics. I was shocked to hear of an experiment done at the big accelerator at the National Accelerator Laboratory, where a person used deuterium. In order to compare his heavy hydrogen results to what might happen with light hydrogen, he had to use data from someone else's experiment on light hydrogen, which was done on different apparatus. When asked why, he said it was because he couldn't get time on the program (because there's so little time and it's such expensive apparatus) to do the experiment with light hydrogen on this apparatus because there wouldn't be any new result. And so the men in charge of programs at NAL are so anxious for new results, in order to get more money to keep the thing going for public relations purposes, they are destroying—possibly—the value of the experiments themselves, which is the whole purpose of the thing. It is often hard for the experimenters there to complete their work as their scientific integrity demands.

All experiments in psychology are not of this type, however. For example, there have been many experiments running rats through all kinds of mazes, and so on—with little clear result. But in 1937 a man named Young did a very interesting one. He had a long corridor with doors all along one side where the rats came in, and doors along the other side where the food was. He wanted to see if he could train the rats to go in at the third door down from wherever he started them off. No. The rats went immediately to the door where the food had been the time before.

Cargo Cult Science: The 1974 Caltech Commencement Address

The question was, how did the rats know, because the corridor was so beautifully built and so uniform, that this was the same door as before? Obviously there was something about the door that was different from the other doors. So he painted the doors very carefully, arranging the textures on the faces of the doors exactly the same. Still the rats could tell. Then he thought maybe the rats were smelling the food, so he used chemicals to change the smell after each run. Still the rats could tell. Then he realized the rats might be able to tell by seeing the lights and the arrangement in the laboratory like any commonsense person. So he covered the corridor, and still the rats could tell.

He finally found that they could tell by the way the floor sounded when they ran over it. And he could only fix that by putting his corridor in sand. So he covered one after another of all possible clues and finally was able to fool the rats so that they had to learn to go in the third door. If he relaxed any of his conditions, the rats could tell.

Now, from a scientific standpoint, that is an A-Number-1 experiment. That is the experiment that makes rat-running experiments sensible, because it uncovers the clues that the rat is really using—not what you think it's using. And that is the experiment that tells exactly what conditions you have to use in order to be careful and control everything in an experiment with rat-running.

I looked into the subsequent history of this research. The next experiment, and the one after that, never referred to Mr. Young. They never used any of his criteria of putting the corridor on sand, or being very careful. They just went right on running rats in the same old way, and paid no attention to the great discoveries of Mr. Young, and his papers are not referred to, because he didn't discover anything about the rats. In fact, he discovered *all* the things you have to do to discover something about rats. But not paying attention to experiments like that is a characteristic of Cargo Cult Science.

Another example is the ESP experiments of Mr. Rhine, and other people. As various people have made criticisms—and they themselves have made criticisms of their own experiments—they improve the techniques so that the effects are smaller, and smaller, and smaller until they gradually disappear. All the parapsychologists are looking for some experiment that can be repeated—that you can do again and get the same effect—statistically, even. They run a million rats—no, it's people this time—they do a lot of things and get a certain statistical effect. Next time they try it they don't get it anymore. And now you find a man saying that it is an irrelevant demand to expect a repeatable experiment. This is *science*?

This man also speaks about a new institution, in a talk in which he was resigning as Director of the Institute of Parapsychology. And, in telling people what to do next, he says that one of the things they have to do is be sure they only train students who have shown their ability to get PSI results to an acceptable extent—not to waste their time on those ambitious and interested students who get only chance results. It is very dangerous to have such a policy in teaching—to teach students only how to get certain results, rather than how to do an experiment with scientific integrity.

So I wish to you—I have no more time, so I have just one wish for you—the good luck to be somewhere where you are free to maintain the kind of integrity I have described, and where you do not feel forced by a need to maintain your position in the organization, or financial support, or so on, to lose your integrity. May you have that freedom. May I also give you one last bit of advice: Never say that you'll give a talk unless you know clearly what you're going to talk about and more or less what you're going to say.

11

It's as Simple as One, Two, Three

✦

An uproarious tale of Feynman the precocious student experiment-ing—with himself, his socks, his typewriter, and his fellow students—to solve the mysteries of counting and of time.

When I was a kid growing up in Far Rockaway, I had a friend named Bernie Walker. We both had "labs" at home, and we would do various "experiments." One time, we were discussing something—we must have been eleven or twelve at the time—and I said, "But thinking is nothing but talking to yourself inside."

"Oh, yeah?" Bernie said. "Do you know the crazy shape of the crankshaft in a car?"

"Yeah, what of it?"

"Good. Now, tell me: How did you describe it when you were talking to yourself?"

So I learned from Bernie that thoughts can be visual as well as verbal.

Later on, in college, I became interested in dreams. I won-dered how things could look so real, just as if light were hit-ting the retina of the eye, while the eyes are closed: Are the

The Pleasure of Finding Things Out

nerve cells on the retina actually being stimulated in some other way—by the brain itself, perhaps—or does the brain have a "judgment department" that gets slopped up during dreaming? I never got satisfactory answers to such questions from psychology, even though I became very interested in how the brain works. Instead, there was all this business about interpreting dreams, and so on.

When I was in graduate school at Princeton a kind of dumb psychology paper came out that stirred up a lot of discussion. The author had decided that the thing controlling the "time sense" in the brain is a chemical reaction involving iron. I thought to myself, "Now, how the hell could he figure that?"

Well, the way he did it was, his wife had a chronic fever which went up and down a lot. Somehow he got the idea to test her sense of time. He had her count seconds to herself (without looking at a clock), and checked how long it took her to count up to 60. He had her counting—the poor woman—all during the day: When her fever went up, he found she counted quicker; when her fever went down, she counted slower. Therefore, he thought, the thing that governed the "time sense" in the brain must be running faster when she's got fever than when she hasn't got fever.

Being a very "scientific" guy, the psychologist knew that the rate of a chemical reaction varies with the surrounding temperature by a certain formula that depends on the energy of the reaction. He measured the differences in speed of his wife's counting, and determined how much the temperature changed the speed. Then he tried to find a chemical reaction whose rates varied with temperature in the same amounts as his wife's counting did. He found that iron reactions fit the pattern best. So he deduced that his wife's sense of time was governed by a chemical reaction in her body involving iron.

It's as Simple as One, Two, Three

Well, it all seemed like a lot of baloney to me—there were so many things that could go wrong in his long chain of reasoning. But it *was* an interesting question: What *does* determine the "time sense"? When you're trying to count at an even rate, what does that rate depend on? And what could you do to yourself to change it?

I decided to investigate. I started by counting seconds—without looking at a clock, of course—up to 60 in a slow, steady rhythm: 1, 2, 3, 4, 5. . . . When I got to 60, only 48 seconds had gone by, but that didn't bother me: The problem was not to count for exactly one minute, but to count at a standard rate. The next time I counted to 60, 49 seconds had passed. The next time, 48. Then 47, 48, 49, 48, 49. . . . So I found I could count at a pretty standard rate.

Now, if I just sat there, without counting, and waited until I thought a minute had gone by, it was very irregular—complete variations. So I found it's very poor to estimate a minute by sheer guessing. But by counting, I could get very accurate.

Now that I knew I could count at a standard rate, the next question was—what affects the rate?

Maybe it has something to do with the heart rate. So I began to run up and down the stairs, up and down, to get my heart beating fast. Then I'd run into my room, throw myself down on the bed, and count up to 60.

I also tried running up and down the stairs and counting to myself *while* I was running up and down.

The other guys saw me running up and down the stairs, and laughed. "What are you doing?"

I couldn't answer them—which made me realize I couldn't talk while I was counting to myself—and kept right on running up and down the stairs, looking like an idiot.

(The guys at the graduate college were used to me looking like an idiot. On another occasion, for example, a guy came into my room—I had forgotten to lock the door during the "experiment"—and found me in a chair wearing my heavy sheepskin coat, leaning out of the wide-open window in the dead of winter, holding a pot in one hand and stirring with the other. "Don't bother me! Don't bother me!" I said. I was stirring Jell-O and watching it closely: I had gotten curious as to whether Jell-O would coagulate in the cold if you kept it moving all the time.)

Anyway, after trying every combination of running up and down the stairs and lying on the bed, surprise! The heart rate had no effect. And since I got very hot running up and down the stairs, I figured temperature had nothing to do with it either (although I must have known that your temperature doesn't really go up when you exercise). In fact, I couldn't find anything that affected my rate of counting.

Running up and down stairs got pretty boring, so I started counting while I did things I had to do anyway. For instance, when I put out the laundry, I had to fill out a form saying how many shirts I had, how many pants, and so on. I found I could write down "3" in front of "pants" or "4" in front of "shirts," but I couldn't count my socks. There were too many of them: I'm already using my "counting machine"—36, 37, 38—and here are all these socks in front of me—39, 40, 41. . . . How do I count the socks?

I found I could arrange them in geometrical patterns—like a square, for example: a pair of socks in this corner, a pair in that one; a pair over here, and a pair over there—eight socks.

I continued this game of counting by patterns, and found I could count the lines in a newspaper article by grouping the lines into patterns of 3, 3, 3, and 1 to get 10; then 3 of those patterns, 3 of those patterns, 3 of those patterns, and 1 of those patterns made 100. I went right down the newspaper like that. After I had finished counting up to 60, I knew where I was in the patterns and could say, "I'm up to 60, and there are 113 lines." I found that I could even *read* the articles while I counted to 60, and it didn't affect the rate! In fact, I could do anything while counting to myself—except talk out loud, of course.

What about typing—copying words out of a book? I found that I could do that, too, but here my time was affected. I was excited: Finally, I've found something that appears to affect my counting rate! I investigated it more.

I would go along, typing the simple words rather fast, counting to myself 19, 20, 21, typing along, counting 27, 28, 29, typing along, until—What the hell is that word? Oh, yeah—and then continue counting 30, 31, 32, and so on. When I'd get to 60, I'd be late.

After some introspection and further observation, I realized what must have happened: I would interrupt my counting when I got to a difficult word that "needed more brains," so to speak. My counting rate wasn't slowing down; rather, the counting itself was being held up temporarily from time to time. Counting to 60 had become so automatic that I didn't even notice the interruptions at first.

The next morning, over breakfast, I reported the results of all these experiments to the other guys at the table. I told them all the things I could do while counting to myself, and said the only thing I absolutely could not do while counting to myself was talk.

One of the guys, a fella named John Tukey, said, "I don't believe you can read, and I don't see why you can't talk. I'll bet you I can talk while counting to myself, and I'll bet you you can't read."

So I gave a demonstration: They gave me a book and I read it for a while, counting to myself. When I reached 60 I said,

"Now!"—48 seconds, my regular time. Then I told them what I had read.

Tukey was amazed. After we checked him a few times to see what his regular time was, he started talking: "Mary had a little lamb; I can say anything I want to, it doesn't make any difference; I don't know what's bothering you"—blah, blah, blah, and finally, "Okay!" He hit his time right on the nose! I couldn't believe it!

We talked about it awhile, and we discovered something. It turned out that Tukey was counting in a different way: He was visualizing a tape with numbers on it going by. He would say, "Mary had a little lamb," and he would *watch* it! Well, now it was clear: He's "looking" at his tape going by, so he can't read, and I'm "talking" to myself when I'm counting, so I can't speak!

After that discovery, I tried to figure out a way of reading out loud while counting—something neither of us could do. I figured I'd have to use a part of my brain that wouldn't interfere with the seeing or speaking departments, so I decided to use my fingers, since that involved the sense of touch.

I soon succeeded in counting with my fingers and reading out loud. But I wanted the whole process to be mental, and not rely on any physical activity. So I tried to imagine the feeling of my fingers moving while I was reading out loud.

I never succeeded. I figured that was because I hadn't practiced enough, but it might be impossible: I've never met anybody who can do it.

By that experience Tukey and I discovered that what goes on in different people's heads when they *think* they're doing the same thing—something as simple as *counting*—is different for different people. And we discovered that you can externally and objectively test how the brain works: You don't have to ask a person how he counts and rely on his own observations of himself; instead, you observe what he can and can't do while he counts. The test is absolute. There's no way to beat it; no way to fake it.

It's natural to explain an idea in terms of what you already have in your head. Concepts are piled on top of each other: This idea is taught in terms of that idea, and that idea is taught in terms of another idea, which comes from counting, which can be so different for different people!

I often think about that, especially when I'm teaching some esoteric technique such as integrating Bessel functions. When I see equations, I see the letters in colors—I don't know why. As I'm talking, I see vague pictures of Bessel functions from Jahnke and Emde's book, with light-tan js, slightly violet-bluish ns, and dark brown xs flying around. And I wonder what the hell it must look like to the students.

12

Richard Feynman
Builds a Universe

✦

*In a previously unpublished interview made under the auspices of the
American Association for the Advancement of Science, Feynman
reminisces about his life in science: his terrifying first lecture to a
Nobel laureate-packed room; the invitation to work on the first
atomic bomb and his reaction; cargo-cult science; and that fateful
predawn wake-up call from a journalist informing him that he'd just
won the Nobel prize. Feynman's answer: "You could have told me
that in the morning."*

NARRATOR:

Mel Feynman was a salesman for a uniform company in
New York City. On May 11, 1918, he welcomed the birth of
his son Richard. Forty-seven years later, Richard Feynman received the Nobel Prize for Physics. In many ways, Mel Feynman had a lot to do with that accomplishment, as Richard
Feynman relates.

FEYNMAN:

Well, before I was born, he [my father] said to my mother
that "this boy is going to be a scientist." You can't say things
like that in front of women's lib these days, but that is what

they said in those days. But he never told me to be a scientist....I learned to appreciate things I had known. There was never any pressure....Later when I got older, he'd take me for walks in the woods and show me the animals and birds and so on...tell me about the stars and the atoms and everything else. He'd tell me what it was about them that was so interesting. He had an attitude about the world and the way to look at it which I found was deeply scientific for a man who had no direct scientific training.

NARRATOR:

Richard Feynman is now professor of physics at the California Institute of Technology in Pasadena, where he has been since 1950. Part of his time he spends teaching and another part he devotes to theorizing about the tiny fragments of matter from which our universe is built. Throughout his career, his sometimes poetic imagination has carried him into many exotic areas: the mathematics involved in creating an atomic bomb, the genetics of a simple virus, and the properties of helium at extremely low temperatures. His Nobel prize–winning work toward developing the theory of quantum electrodynamics helped solve many physical problems more directly and more efficiently than had ever been possible. But again, what set that long train of accomplishments in motion were long walks in the woods with his father.

FEYNMAN:

He had ways of looking at things. He used to say, "Suppose we were Martians and we came down to the earth and then we would see these strange creatures doing things; what would we think? For instance," he would say, "to take an example, suppose that we never went to sleep. We're Martians, but we have a consciousness that works all the time, and we find these creatures who for eight hours every day stop and close their eyes and become more or less inert. We'd have an interesting question to ask them. We'd say, 'How does it feel to do that all the time? What happens to your ideas? You're running along very well, you're thinking clearly—and what happens? Do they suddenly stop? Or do they go more and more slowly and stop, or exactly how do you turn off thoughts?'" Then later I thought about that a lot and I did experiments when I was in college to try to find out the answer to that—what happened to your thoughts when you went to sleep.

NARRATOR:

In his early days, Dr. Feynman planned to be an electrical engineer, to get his hands into physics and make it do useful things for him and the world around him. It didn't take him long to realize that he was really more interested in what made things work, in the theoretical and mathematical principles that underlie the operation of the universe itself. His mind became his laboratory.

FEYNMAN:

When I was young, what I call the laboratory was just a place to fiddle around, make radios and gadgets and photocells and whatnot. I was very shocked when I discovered what they call a laboratory in a university. That's a place where you are supposed to measure something very seriously. I never measured a damn thing in my laboratory. I just fiddled around and made things. That was the kind of lab I had when I was young and I thought entirely that way. I thought that was the way I was going to go. Well, in that lab, I had to solve certain problems. I used to repair radios. I had to, for example, get some resistance to put in line with some voltmeters so it would run in different scales. Things like that. So I began to find these formulas, electrical formulas, and a friend of mine had a book with electrical formulas in it and [it] had relations between the resistors. It had things like, the power is

The Pleasure of Finding Things Out

the square of the current times the voltage. The voltage divided by the current is the resistance and all; it had six or seven formulas. It seemed to me that they were all related, they really weren't all independent, that one could come from the other. And so, I got to fiddling about and I understood from the algebra I had been learning in school how to do it. I realized that mathematics was somehow important in this business.

So I got more and more interested in the mathematical business associated with physics. In addition, mathematics by itself had a great appeal for me. I loved it all my life. [...]

NARRATOR:
After graduation from the Massachusetts Institute of Technology, Richard Feynman moved approximately 400 miles southwest to Princeton University, where he would eventually get his Ph.D. It was there, at the age of 24 that he gave his first formal lecture. It was a very eventful lecture, as it turned out.

FEYNMAN:
When I was an undergraduate I worked with Professor Wheeler* as a research assistant, and we had worked out together a new theory about how light worked, how the interaction between atoms in different places worked; and it was at that time an apparently interesting theory. So Professor Wigner†, who was in charge of the seminars there, suggested that we give a seminar on it, and Professor Wheeler said that

---

*John Archibald Wheeler (1911–), physicist, best known to the general public for having coined the term "black hole." *Ed.*

†Eugene P. Wigner (1902–1995), 1963 Nobel Prize in Physics, for his contributions to the theory of the atomic nucleus and elementary particles, through his work on symmetry principles. *Ed.*

since I was a young man and hadn't given seminars before, it would be a good opportunity to learn how to do it. So this was the first technical talk that I ever gave.

I started to prepare the thing. Then Wigner came to me and said that he thought the work was important enough that he'd made special invitations to the seminar to Professor Pauli, who was a great professor of physics visiting from Zurich; to Professor von Neumann, the world's greatest mathematician; to Henry Norris Russell, the famous astronomer; and to Albert Einstein, who was living near there. I must have turned absolutely white or something because he said to me, "Now don't get nervous about it, don't be worried about it. First of all, if Professor Russell falls asleep, don't feel bad, because he always falls asleep at lectures. When Professor Pauli nods as you go along, don't feel good, because he always nods, he has palsy," and so on. That kind of calmed me down a bit, but I was still worried. So Professor Wheeler promised me that he would answer all the questions and all that I would do would be to give the lecture.

So I remember coming in—you can imagine that first time, it was like going through fire. I had written all the equations on the blackboard way ahead of time so that all the blackboards were full of equations. People don't want so many equations . . . they want to understand the ideas better. And then I remember getting up to talk and there were these great men in the audience and it was frightening. And I can still see my own hands as I pulled out the papers from the envelope that I had them in. They were shaking. As soon as I got the paper out and started to talk, something happened to me which has always happened since and which is a wonderful thing. If I'm talking physics, I love the thing, I think only about physics, I don't worry where I am; I don't worry about anything. And everything went very easily. I simply explained the whole business as best I could. I didn't think about who was there. I was thinking only about the problem I was explaining. And then at the end when the question time came, I had nothing to worry about because Professor Wheeler was going to answer them. Professor Pauli stood up—he was sitting next to Professor Einstein. He said, "I do not think this theory can be right because of this and this and that and the other thing and so forth, don't you agree, Professor Einstein?" Einstein said, "No-o-o-o," and that was the nicest no I ever heard.

NARRATOR:

It was at Princeton that Richard Feynman learned that even if he lived his entire life in the world of mathematics and theoretical physics, there was another world out there that would insist on making very practical demands of him. In those years the world was at war, and the United States had just begun to work on the atomic bomb.

FEYNMAN:

Just about that time, Bob Wilson came into my room to tell me about a project he was starting that had to do with making uranium for atomic bombs. He said there was a meeting at 3:00 and it was a secret, but he knew that when I knew what the secret was I would have to go along with it, so there was no harm in telling me. I said, "You made a mistake in telling me the secret. I am not going along with you. I'm just going back and do my work—back to working on my thesis." He went out of the room saying "We're going to have a meeting at 3:00." That [happened] in the morning. I started to pace the floor and think about the consequences of the bomb being in the hands of the Germans and all that stuff and decided that it was very exciting and important to do. So I was at the meeting at 3:00 and I stopped working on my degree.

The problem was that you had to separate the isotopes of uranium in order to make a bomb. The uranium came in two isotopes and U_{235} was the reactive one and you wanted to separate them. Wilson had invented a scheme for doing the separation—making a beam of ions and bunching the ions—the velocity of the two isotopes at the same energy is slightly different. So if you make little lumps and they go down a long tube, one gets ahead of the other and you can separate them that way. That's the plan he had. I was theoretical by that time. What I was originally set to do was find out if the device as it was designed was at all practical; could it be done at all? There were a lot of questions about space charge limitations and so on and I deduced that it could be done.

NARRATOR:

Even though Feynman deduced that Wilson's method for separating uranium isotopes was indeed theoretically possible, another method was eventually used to produce uranium-235 for the atomic bomb. Nevertheless, there was still plenty for Richard Feynman and his high-level theorizing to do at the main laboratory in Los Alamos, New Mexico, charged with developing the bomb. After the war, he joined the staff of the Laboratory of Nuclear Studies at Cornell University. Today, he has mixed emotions about the work he did toward making the atomic bomb possible. Had he done the right thing or the wrong thing?

FEYNMAN:

No, I don't think that I was wrong exactly at the time I made the decision. I thought about it and I think correctly that it was very dangerous if the Nazis got it. There was, however, I think, an error in my thought in that after the Germans were defeated—that was much later, three or four years later—we were working very hard. I didn't stop; I didn't even consider that the motive for originally doing it was no longer there. And that's one thing I did learn, that if you have some reason for doing something that's very strong and you start working at it, you must look around every once in a while and find out if the original motives are still right. At the time I made the decision, I think that was right, but to continue without thinking about it may have been wrong. I don't know what would have happened if I had thought about it. I may have decided to continue anyway, I don't know. But the point of not thinking about it when the original conditions that made [me make] the original decision had changed, that's a mistake.

NARRATOR:

After five stimulating years at Cornell, Dr. Feynman, like many other easterners before and after him, was lured to California and the equally stimulating surroundings of the California Institute of Technology. And there were other reasons.

FEYNMAN:

First of all, the weather is no good in Ithaca. Secondly, I kind of like going out to nightclubs and stuff like that.

Bob Bacher invited me to come out here to give a series of lectures on some work I had developed at Cornell. So I gave the lecture and then he said, "May I lend you my car?" I enjoyed that and I took his car and every night I went to Hollywood and the Sunset Strip and hung around there and had a good time, and that mixture of good weather and a wider horizon than is available in a small town in upper New York State is what finally convinced me to come here. It wasn't very hard. It wasn't a mistake. There was another decision that wasn't a mistake.

NARRATOR:

On the California Institute of Technology faculty, Dr. Feynman serves as Richard Chace Tolman Professor of Theoretical Physics. In 1954 he received the Albert Einstein Award, and in 1962 the Atomic Energy Commission gave him the E. O. Laurence Award for "especially meritorious contributions to the development, use or control of atomic energy." Finally, in 1965, he received the greatest scientific award of all, the Nobel Prize. He shared it with Sin-Itiro Tomonaga of Japan and Julian Schwinger of Harvard. For Dr. Feynman, the Nobel Prize was a rude awakening.

FEYNMAN:

The telephone rang, the guy said [he was] from some broadcasting company. I was very annoyed to be awakened. That was my natural reaction. You know, you're half awake and you're annoyed. So the guy says, "We'd like to inform you that you've won the Nobel Prize." And I'm thinking to myself—I'm still annoyed, see—it didn't register. So I said, "You could have told me that in the morning." So he says, "I thought you'd like to know." Well, I said I was asleep and put the telephone back. My wife said, "What was that?" and I said, "I won the Nobel Prize." And she said, "Go on, you're kidding me." I've often tried to fool her but I can never fool her. Every time I try to fool her she sees through me, so this time she was wrong. She thought I was kidding. She thought it was some student, some drunken student or something. So she didn't believe me. But when the second telephone call came ten minutes later from another newspaper, I said to that fellow, "Yes, I've already heard it, leave me alone." Then I took the receiver off the hook and I thought I'd just go back to sleep and by 8:00 I'd put the receiver back on the hook. I couldn't go back to sleep and my wife couldn't either. I got up and walked around, and finally I put the receiver back and started to answer the phone.

Some short time after that, I had a ride in a taxi somewhere and the taxi driver is talking and I'm talking and I'm telling him my problems about how these guys ask me and I don't know how to explain it. He says, "I heard an interview of yours. I saw it on TV. The guy says to you, 'Would you please explain what you did to get the prize in two minutes.' And you tried to do it and you're crazy. You know what I would have said? 'Hell, man, if I could have told you in two minutes, I wouldn't be worth the Nobel Prize.'" So that's the answer I gave after that. When somebody asks me, I always tell them, look, if I could explain it that easily, it wouldn't have been worth the Nobel Prize. It's not really fair, but it's kind of a fun answer.

NARRATOR:

As mentioned earlier, Dr. Feynman received the Nobel Prize for his contributions to developing a theory that would define the newly emerging field of quantum electrodynamics. It is, as Dr. Feynman puts it, "the theory of everything else." It does not apply to nuclear energy or the forces of gravity, but it does apply to the interaction of electrons with particles of light called photons. It underlies the way electricity flows, the phenomenon of magnetism, and the way that X-rays are produced and interact with other forms of matter. The "quantum" in quantum electrodynamics recognizes a theory of the mid-twenties which states that the electrons surrounding the nucleus of every atom are limited to certain quantum states or energy levels. They can exist only at those levels and nowhere in between. These quantized energy levels are determined by the intensity of the light that falls on the atom, among other things.

FEYNMAN:

One of the biggest and most important tools of theoretical physics is the wastebasket. You have to know when to leave it alone, hmmm? In fact, I learned almost everything I know about electricity, magnetism, and quantum mechanics and everything else in attempting to develop that theory. And what I got a Nobel Prize for ultimately was that in 1947, the regular people's theory, the ordinary theory which I was trying to fix by changing it, was in some trouble so I was trying to fix it, but Bethe had found out that if you do just the right things, if you kind of forget some things and don't forget other things, do it just right, you can get the right answers compared to experiment and he made some suggestions to me. And I knew so much about electrodynamics by this time from having tried this crazy theory and written it in some 655 different forms that I knew how to do what he wanted, how to control and organize this calculation in a very smooth and convenient way and have powerful methods to do it. In other words, I used the stuff, the machinery which I had developed to evolve my own theory on the old theory—sounds like the obvious thing to do, but I didn't think of it for years—and found out it was extremely powerful by that time and I could do things by the old theory much faster than anyone else had before.

NARRATOR:

In addition to a lot of other things, Dr. Feynman's theory of quantum electrodynamics provides new insights into understanding the forces that hold matter together. It also adds a little bit more to what we know about the properties of the infinitesimally small, short-lived particles from which everything else in the universe is composed. As physicists have probed deeper and deeper into the structure of nature, they have found that what once seemed very simple may be very complex and what once seemed very complex may be very simple. Their tools are the high-powered atom smashers that can fracture atomic particles into smaller and smaller fragments.

FEYNMAN:

When we start out, we look at matter and we see many different phenomena—winds and waves and moon and all that kind of stuff. And we try to reorganize it. Is the motion of the wind like the motion of the waves and so forth? Gradually we find that many, many things are similar. It's not as big a variety as we think. We get all the phenomena and we get the principles underneath, and one of the most useful principles seemed to be the idea that things are made of other things. We found, for example, that all matter was made out of atoms, and then a large amount is understood as long as you understand the properties of atoms. And at first the atoms are supposed to be simple, but it turns out that in order to explain all the varieties, the phenomena of matter, the atoms have to be more complicated, and that there are 92 atoms. In fact, there are many more, because they have different weights. Then to understand the variety of the properties of atoms is the next problem. And we find that we can understand that if we make out that the atoms themselves are made of constituents—in this particular case, the nucleus around which the electrons go. And that all the different atoms are just different numbers of electrons. It's a beautifully unifying system that works.

All the different atoms are just the same thing with different numbers of electrons. However, the nuclei then differ. And so we start to study the nuclei. And there was a great variety as soon as we started experiments hitting nuclei together—Rutherford and so forth. From 1914 on, they discovered that they were complicated at first. But then it was realized that they could be understood if they're made of constituents, too. They are made of protons and neutrons. And they interact with some force that holds them together. In order to understand the nuclei, we have to understand that force a little bit better. Incidentally, in the case of atoms there was also a force; that's an electrical force and that we understood. So besides electrons there was also the electrical force, which we represent by photons of light. The light and electrical force are integrated into one thing called photons, so the outside world, so to speak, outside the nucleus is electrons and photons. And the theory of the behavior of electrons is quantum electrodynamics and that's what I got the Nobel Prize for working on.

But now we go into the nucleus and find that they could be made of protons and neutrons, but there's this strange force. Trying to understand that force is the next problem. And various suggestions that there might be other particles were made by Yukawa,* and so we did experiments hitting the protons and neutrons together with higher energy and indeed new things came out, just like when you hit electrons together with high enough energy, photons came out. So we have these new things coming out. They were mesons. So it looked like Yukawa was right. We continued the experiment. And then what happened to us was that we got a tremendous variety of particles; not just one kind of photon, you see, but we hit photons and neutrons together and we got over 400 different kinds of particles—lambda particles and sigma particles. They're all different. And π mesons and K mesons and so on. Well, we incidentally also made muons, but they have nothing to do with neutrons and protons apparently. At least no more than electrons do. That's a strange extra part that we don't understand where it goes. It's just like an electron but heavier. So we have electrons and muons out here which don't interact strongly with these other things. These other things we call strongly interacting particles, or hadrons. And they include protons and neutrons and all the things you get immediately when you hit them together very hard. So now the problem is to try and represent the properties of all these particles in some organizing fashion. And that's a great game and we're all working on it. It's called high-energy physics or fundamental particle physics. It used to be called fundamen-

---

*Hideki Yukawa (1907–1981), winner of the 1949 Nobel Prize in Physics for predicting the existence of mesons. *Ed.*

tal particle physics, but nobody can believe that 400 different constituents are fundamental. Another possibility is that they are themselves made of some deeper constituent. And that seems to be a reasonable possibility. And so it turns out that a theory has been invented—the theory of quarks; that certain of these things like the proton, for instance, or neutron, are made of three objects called quarks.

NARRATOR:

No one has yet seen a quark, which is too bad, because they may represent the fundamental building block for all the other more complicated atoms and molecules that make up the universe. The name was chosen for no particular reason by Dr. Feynman's colleague, Murray Gell-Mann, some years ago. Somewhat to Dr. Gell-Mann's surprise, the Irish novel writer James Joyce had already anticipated that name thirty years earlier in his book, *Finnegan's Wake*. The key phrase was "three quarks for Muster Mark." This was even a bigger coincidence since, as Dr. Feynman explained, the quarks that make up the particles of the universe seem to come in threes. In the search for quarks, physicists are knocking protons and neutrons together at such high energies with the hope that they will break apart into their quark components in the process.

FEYNMAN:

All true, and one of the things that's been holding up the quark theory was that it's obviously cockeyed, because if the things were made of quarks, if we hit two protons together, we ought to produce three quarks sometimes. It turns out that in this quark model that we are talking about, the quarks carry very peculiar electrical charges. All the particles in the world that we know contain integral charges. Usually one electric charge plus or minus or nothing. But the theory of quarks has it that the quarks carry charges like minus a third or plus two thirds of an electric charge. And if such a particle exists, it would be obvious, because the number of bubbles it would leave in a bubble chamber when it made a track would be much [smaller]. Say you had a charge of a third; then it whips up one ninth as many atoms as it turns out—the square—along the track, so there would be one ninth as many bubbles along the track as you would get for an ordinary particle. And that's obvious; if you see a lightly drawn track, there's something wrong. And they've looked and looked for such a track, and they haven't found them yet. So that's one of the serious problems. That's the excitement. Are we on the right track or are we walking around in the utter darkness when the answer is way over here to the right, or are we smelling it closely and just haven't quite got it right? And if we just get it right, we'll suddenly understand why that experiment looks different.

NARRATOR:

And what if these high-powered experiments with atom smashers and bubble chambers do show that the world is made of quarks? Will we ever be able to see them in a practical way?

FEYNMAN:

Well, for the problem of understanding the hadrons and the muons and so on, I can see at the present time no practical applications at all, or virtually none. In the past many people have said that they could see no applications and then later they found applications. Many people would promise under those circumstances that something's bound to be useful. However, to be honest—I mean he looks foolish; saying there will never be anything useful is obviously a foolish thing to do. So I'm going to be foolish and say these damn things will never have any application, as far as I can tell. I'm too dumb to see it. All right? So why do you do it? Applications aren't the only thing in the world. It's interesting in understanding what the world is made of. It's the same interest, the curiosity of man that makes him build telescopes. What is the use of discovering the age of the universe? Or what are these quasars that are exploding at long distances? I mean what's the use of all that astronomy? There isn't any. Nonetheless, it's interesting. So it's the same kind of exploration of our world that I'm following and it's curiosity that I'm satisfying. If human curiosity represents a need, the attempt to satisfy curiosity, then this is practical in the sense that it is that. That's the way I would look at it at the present time. I would not put out any promise that it would be practical in some economic sense.

NARRATOR:

As for science itself and what it means to all of us, Dr. Feynman says he is reluctant to philosophize on the subject. Nevertheless, that does not prevent him from coming up with some interesting and provocative ideas about what he believes science is and what it is not.

FEYNMAN:

Well, I'll say it is the same as it always was from the day it began. It's the pursuit of understanding of some subject or some thing based on the principle that what happens in nature is true and is the judge of the validity of any theory about it. If Lysenko says that you cut off rats' tails for 500 generations, then the new rats that are born will not have tails. (I don't know if he says that or not. Let's say Mr. Jones says that.) Then if you try it and it doesn't work, we know that it isn't true. That principle, the separation of the true from the false by experiment or experience, that principle and the resultant body of knowledge which is consistent with that principle, that is science.

To science we also bring, besides the experiment, a tremendous amount of human intellectual attempt at generalization.

So it's not merely a collection of all those things which just happen to be true in experiments. It's not just a collection of facts about what happens when you cut off [rats'] tails because it would be much too much for us to hold in our heads. We've found a large number of generalizations. For example, if it's true of rats and cats, we say it's true of mammals. Then we discover if it's true of other animals; then we discover it's true of plants, and finally it becomes a property of life to a certain extent that we don't inherit as an acquired characteristic. It's not exactly true, actually, absolutely. We later discovered experiments that show that cells can carry information through the mitochondria or something so that we modify it as we go along. But as all the principles must be as wide as possible, must be as general as possible, and still be in complete accord with experiment, that's the challenge.

You see, the problem of obtaining facts from experience—it sounds very, very simple. You just try it and see. But man is a weak character and it turns out to be much more difficult than you think to just try it and see. For instance, you take education. Some guy comes along and he sees the way people teach mathematics. And he says, "I have a better idea. I'll make a toy computer and teach them with it." So he tries it with a group of children, he hasn't got a lot of children, maybe somebody gives him a class to try it with. He loves what he's doing. He's excited. He understands completely what his thing is. The kids know that it's something new, so they're all excited. They learn very, very well and they learn the regular arithmetic better than the other kids did. So you make a test—they learn arithmetic. Then this is registered as a fact—that the teaching of arithmetic can be improved by this method. But it's not a fact, because one of the conditions of the experiment was that the particular man who invented it was doing the teaching. What you really want to know is, if you just had this method described in a book to an average teacher (and you have to have average teachers; there are teachers all over the world and there must be many who are average), who then gets this book then tries to teach it with the method described, will it be better or not? In other words, what happens is that you get all kinds of statements of fact about education, about sociology, even psychology—all kinds of things which are, I'd say, pseudoscience. They've done statistics which they say they've done very carefully. They've done experiments which are not really controlled experiments. [The results] aren't really repeatable in controlled experiments. And they report all this stuff. Because science which is done carefully has been successful; by doing something like that, they think that they get some honor. I have an example.

In the Solomon Islands, as many people know, the natives didn't understand the airplanes which came down during the war and brought all kinds of goodies for the soldiers. So now they have airplane cults. They make artificial landing strips and they make fires along the landing strips to imitate the lights and this poor native sits in a wooden box he's built with wooden earphones with bamboo sticks going up to represent the antenna and turning his head back and forth, and they have radar domes made of wood and all kinds of things hoping to lure the airplanes to give goods to them. They're imitating the action. It's just what the other guy did. Well, a hell of a lot of our modern activity in many, many fields is that kind of science. Just like aviation. That's a science. The science of education, for example, is no science at all. It's a lot of work. It takes a lot of work to carve those things out, those wooden airplanes. But it doesn't mean that they are actually finding out something. Penology, prison reform—to understand why people do crimes; look at the world—we understand it more and more with our modern understanding of these things. More about education, more about crime; the scores on the tests are going down and there's more people in prison; young people are committing crimes, we just don't understand it at all. It just isn't working, to discover things about these things by using the scientific method in the type of imitation which they are using now. Now whether the scientific method would work in these fields if we knew how to do it, I don't know. It's particularly weak in this way. There may be some other method. For example, to listen to the ideas of the past and the experience of people for a long time might be a good idea. It's only a good idea not to pay attention to the past when you have another independent source of information that you've decided to follow. But you've got to watch out who it is you're following if you're going to [ignore] the wisdom of the people who've looked at this thing and thought about it and unscientifically came to a conclusion. They have no less right to be right than you have to be right in modern times; to equally unscientifically come to a conclusion.

Well, how's that? Am I doing okay as a philosopher?

NARRATOR:

In this edition of the Future for Science—a taped series of interviews with Nobel laureates—you've heard Dr. Richard Feynman of the California Institute of Technology. The series has been prepared under the auspices of the American Association for the Advancement of Science.

13

The Relation
of Science and Religion

In a kind of thought experiment, Feynman takes the various points of view of an imaginary panel to represent the thinking of scientists and spiritualists and discusses the points of agreement and of disagreement between science and religion, anticipating, by two decades, the current active debate between these two fundamentally different ways of searching for truth. Among other questions, he wonders whether atheists can have morals based on what science tells them, in the way that spiritualists can have morals based on their belief in God—an unusually philosophical topic for pragmatic Feynman.

In this age of specialization, men who thoroughly know one field are often incompetent to discuss another. The great problems of the relations between one and another aspect of human activity have for this reason been discussed less and less in public. When we look at the past great debates on these subjects, we feel jealous of those times, for we should have liked the excitement of such argument. The old problems, such as the relation of science and religion, are still with us, and I believe present as difficult dilemmas as ever, but they are not often publicly discussed because of the limitations of specialization.

But I have been interested in this problem for a long time and would like to discuss it. In view of my very evident lack of knowledge and understanding of religion (a lack which will grow more apparent as we proceed), I will organize the discussion in this way: I will suppose that not one man but a group of men are discussing the problem, that the group consists of specialists in many fields—the various sciences, the various religions and so on—and that we are going to discuss the problem from various sides, like a panel. Each is to give his point of view, which may be molded and modified by the later discussion. Further, I imagine that someone has been chosen by lot to be the first to present his views, and I am he so chosen.

I would start by presenting the panel with a problem: A young man, brought up in a religious family, studies a science, and as a result he comes to doubt—and perhaps later to disbelieve in—his father's God. Now, this is not an isolated example; it happens time and time again. Although I have no statistics on this, I believe that many scientists—in fact, I actually believe that more than half of the scientists—really disbelieve in their father's God; that is, they don't believe in a God in a conventional sense.

Now, since the belief in a God is a central feature of religion, this problem that I have selected points up most strongly the problem of the relation of science and religion. Why does this young man come to disbelieve?

The first answer we might hear is very simple: You see, he is taught by scientists, and (as I have just pointed out) they are all atheists at heart, so the evil is spread from one to another. But if you can entertain this view, I think you know less of science than I know of religion.

Another answer may be that a little knowledge is dangerous; this young man has learned a little bit and thinks he knows it all, but soon he will grow out of this sophomoric sophistication and come to realize that the world is more complicated, and he will begin again to understand that there must be a God.

I don't think it is necessary that he come out of it. There are many scientists—men who hope to call themselves mature—who still don't believe in God. In fact, as I would like to explain later, the answer is not that the young man thinks he knows it all—it is the exact opposite.

A third answer you might get is that this young man really doesn't understand science correctly. I do not believe that science can disprove the existence of God; I think that is impossible. And if it is impossible, is not a belief in science and in a God—an ordinary God of religion—a consistent possibility?

Yes, it is consistent. Despite the fact that I said that more than half of the scientists don't believe in God, many scientists *do* believe in both science and God, in a perfectly consistent way. But this consistency, although possible, is not easy to attain, and I would like to try to discuss two things: Why it is not easy to attain, and whether it is worth attempting to attain it.

When I say "believe in God," of course, it is always a puzzle—what is God? What I mean is the kind of personal God, characteristic of the Western religions, to whom you pray and who has something to do with creating the universe and guiding you in morals.

For the student, when he learns about science, there are two sources of difficulty in trying to weld science and religion together. The first source of difficulty is this—that it is imperative in science to doubt; it is absolutely necessary, for progress in science, to have uncertainty as a fundamental part of your inner nature. To make progress in understanding, we must remain modest and allow that we do not know. Nothing is certain or proved beyond all doubt. You investigate for curiosity, because it is *unknown,* not because you know the answer. And as you develop more information in the sciences, it is not that you are finding out the truth, but that you are finding out that this or that is more or less likely.

That is, if we investigate further, we find that the statements of science are not of what is true and what is not true, but statements of what is known to different degrees of certainty: "It is very much more likely that so and so is true than that it is not true"; or "such and such is almost certain but there is still a little bit of doubt"; or—at the other extreme—"well, we really don't know." Every one of the concepts of science is on a scale graduated somewhere between, but at neither end of, absolute falsity or absolute truth.

It is necessary, I believe, to accept this idea, not only for science, but also for other things; it is of great value to acknowledge ignorance. It is a fact that when we make decisions in our life, we don't necessarily know that we are making them correctly; we only think that we are doing the best we can—and that is what we should do.

Attitude of Uncertainty

I think that when we know that we actually do live in uncertainty, then we ought to admit it; it is of great value to realize that we do not know the answers to different questions. This attitude of mind—this attitude of uncertainty—is vital to the scientist, and it is this attitude of mind which the student must first acquire. It becomes a habit of thought. Once acquired, one cannot retreat from it anymore.

What happens, then, is that the young man begins to doubt everything because he cannot have it as absolute truth. So the question changes a little bit from "Is there a God?" to "How sure is it that there is a God?" This very subtle change is a great stroke and represents a parting of the ways between science and religion. I do not believe a real scientist can ever believe in the same way again. Although there are scientists who believe in God, I do not believe that they think of God in the same way as religious people do. If they are consistent with their science, I think that they say something like this to themselves: "I am almost certain there is a God. The doubt is very small." That is quite different from saying, "I know that there is a God." I do not believe that a scientist can ever obtain that view—that really religious understanding, that real knowledge that there is a God—that absolute certainty which religious people have.

Of course this process of doubt does not always start by attacking the question of the existence of God. Usually special tenets, such as the question of an afterlife, or details of the religious doctrine, such as details of Christ's life, come under scrutiny first. It is more interesting, however, to go right into the central problem in a frank way, and to discuss the more extreme view which doubts the existence of God.

Once the question has been removed from the absolute, and gets to sliding on the scale of uncertainty, it may end up in very different positions. In many cases it comes out very close to being certain. But on the other hand, for some, the net result of close scrutiny of the theory his father held of God may be the claim that it is almost certainly wrong.

Belief in God—and the Facts of Science

That brings us to the second difficulty our student has in trying to weld science and religion: Why does it often end up that the belief in God—at least, the God of the religious type—is considered to be very unreasonable, very unlikely? I think that the answer has to do with the scientific things—the facts or partial facts—that the man learns.

For instance, the size of the universe is very impressive, with us on a tiny particle whirling around the sun, among a hundred thousand million suns in this galaxy, itself among a billion galaxies.

Again, there is the close relation of biological man to the animals, and of one form of life to another. Man is a late-comer in a vast evolving drama; can the rest be but a scaffolding for his creation?

Yet again, there are the atoms of which all appears to be constructed, following immutable laws. Nothing can escape it; the stars are made of the same stuff, and the animals are made of the same stuff, but in such complexity as to mysteriously appear alive—like man himself.

It is a great adventure to contemplate the universe beyond man, to think of what it means without man—as it was for the great part of its long history, and as it is in the great majority of places. When this objective view is finally attained, and the mystery and majesty of matter are appreciated, to then turn the objective eye back on man viewed as matter, to see life as part of the universal mystery of greatest depth, is to sense an experience which is rarely described. It usually ends in laughter, delight in the futility of trying to understand. These scientific views end in awe and mystery, lost at the edge in uncertainty, but they appear to be so deep and so impressive that the theory that it is all arranged simply as a stage for God to watch man's struggle for good and evil seems to be inadequate.

So let us suppose that this is the case of our particular student, and the conviction grows so that he believes that individual prayer, for example, is not heard. (I am not trying to disprove the reality of God; I am trying to give you some idea of–some sympathy for–the reasons why many come to think that prayer is meaningless.) Of course, as a result of this doubt, the pattern of doubting is turned next to ethical problems, because, in the religion which he learned, moral problems were connected with the word of God, and if the God doesn't exist, what is his word? But rather surprisingly, I think, the moral problems ultimately come out relatively unscathed; at first perhaps the student may decide that a few little things were wrong, but he often reverses his opinion later, and ends with no fundamentally different moral view.

There seems to be a kind of independence in these ideas. In the end, it is possible to doubt the divinity of Christ, and yet to believe firmly that it is a good thing to do unto your neighbor as you would have him do unto you. It is possible to have both these views at the same time; and I would say that I hope you will find that my atheistic scientific colleagues often carry themselves well in society.

Communism and the Scientific Viewpoint

I would like to remark, in passing, since the word "atheism" is so closely connected with "communism," that the communist views are the antithesis of the scientific, in the sense that in communism the answers are given to all the questions–political questions as well as moral ones–without discussion and without doubt. The scientific viewpoint is the exact opposite of this; that is, all questions must be doubted and discussed; we must argue everything out–observe things, check them, and so change them. The democratic government is much closer to this idea, because there is discussion and a chance of modification. One doesn't launch the ship in a definite direction. It is true that if you have a tyranny of ideas, so that you know exactly what has to be true, you act very decisively, and it looks good—for a while. But soon the ship is heading in the wrong direction, and no one can modify the direction anymore. So the uncertainties of life in a democracy are, I think, much more consistent with science.

Although science makes some impact on many religious ideas, it does not affect the moral content. Religion has many aspects; it answers all kinds of questions. First, for example, it answers questions about what things are, where they come from, what man is, what God is—the properties of God, and so on. Let me call this the metaphysical aspect of religion. It also tells us another thing—how to behave. Leave out of this the idea of how to behave in certain ceremonies, and what rites to perform; I mean it tells us how to behave in life in general, in a moral way. It gives answers to moral questions; it gives a moral and ethical code. Let me call this the ethical aspect of religion.

Now, we know that, even with moral values granted, human beings are very weak; they must be reminded of the moral values in order that they may be able to follow their consciences. It is not simply a matter of having a right conscience; it is also a question of maintaining strength to do what you know is right. And it is necessary that religion give strength and comfort and the inspiration to follow these moral views. This is the inspirational aspect of religion. It gives inspiration not only for moral conduct—it gives inspiration for the arts and for all kinds of great thoughts and actions as well.

Interconnections

These three aspects of religion are interconnected, and it is generally felt, in view of this close integration of ideas, that to attack one feature of the system is to attack the whole structure. The three aspects are connected more or less as follows: The moral aspect, the moral code, is the word of God—which involves us in a metaphysical question. Then the inspiration comes because one is working the will of God; one is for God; partly one feels that one is with God. And this is a great inspiration because it brings one's actions in contact with the universe at large.

So these three things are very well interconnected. The difficulty is this: that science occasionally conflicts with the first of the three categories—the metaphysical aspect of religion. For instance, in the past there was an argument about whether the earth was the center of the universe—whether the earth moved around the sun or stayed still. The result of all this was a terrible strife and difficulty, but it was finally resolved—with religion retreating in this particular case. More recently there was a conflict over the question of whether man has animal ancestry.

The result in many of these situations is a retreat of the religious metaphysical view, but nevertheless, there is no collapse of the religion. And further, there seems to be no appreciable or fundamental change in the moral view.

After all, the earth moves around the sun—isn't it best to turn the other cheek? Does it make any difference whether the earth is standing still or moving around the sun? We can expect conflict again. Science is developing and new things will be found out which will be in disagreement with the present-day metaphysical theory of certain religions. In fact, even with all the past retreats of religion, there is still real conflict for particular individuals when they learn about the science and they have heard about the religion. The thing has not been integrated very well; there are real conflicts here—and yet morals are not affected.

The Pleasure of Finding Things Out

As a matter of fact, the conflict is doubly difficult in this metaphysical region. Firstly, the facts may be in conflict, but even if the facts were not in conflict, the attitude is different. The spirit of uncertainty in science is an attitude toward the metaphysical questions that is quite different from the certainty and faith that is demanded in religion. There is definitely a conflict, I believe—both in fact and in spirit—over the metaphysical aspects of religion.

In my opinion, it is not possible for religion to find a set of metaphysical ideas which will be guaranteed not to get into conflicts with an ever-advancing and always-changing science which is going into an unknown. We don't know how to answer the questions; it is impossible to find an answer which someday will not be found to be wrong. The difficulty arises because science and religion are both trying to answer questions in the same realm here.

Science and Moral Questions

On the other hand, I don't believe that a real conflict with science will arise in the ethical aspect, because I believe that moral questions are outside of the scientific realm.

Let me give three or four arguments to show why I believe this. In the first place, there have been conflicts in the past between the scientific and the religious view about the metaphysical aspect and, nevertheless, the older moral views did not collapse, did not change.

Second, there are good men who practice Christian ethics and who do not believe in the divinity of Christ. They find themselves in no inconsistency here.

Thirdly, although I believe that from time to time scientific evidence is found which may be partially interpreted as giving some evidence of some particular aspect of the life of Christ, for example, or of other religious metaphysical ideas, it seems to me that there is no scientific evidence bearing on the Golden Rule. It seems to me that that is somehow different.

Now, let's see if I can make a little philosophical explanation as to why it is different—how science cannot affect the fundamental basis of morals.

The typical human problem, and one whose answer religion aims to supply, is always of the following form: Should I do this? Should we do this? Should the government do this? To answer this question we can resolve it into two parts: First—If I do this, what will happen?—and second—Do I want that to happen? What would come of it of value—of good?

Now a question of the form: If I do this, what will happen? is strictly scientific. As a matter of fact, science can be defined as a method for, and a body of information obtained by, trying to answer only questions which can be put into the form: If I do this, what will happen? The technique of it, fundamentally, is: Try it and see. Then you put together a large amount of information from such experiences. All scientists will agree that a question—any question, philosophical or other—which cannot be put into the form that can be tested by experiment (or, in simple terms, that cannot be put into the form: If I do this, what will happen?) is not a scientific question; it is outside the realm of science.

I claim that whether you want something to happen or not—what value there is in the result, and how you judge the value of the result (which is the other end of the question: Should I do this?), must lie outside of science because it is not a question that you can answer only by knowing what happens; you still have to *judge* what happens—in a moral way. So, for this theoretical reason I think that there is a complete consistency between the moral view—or the ethical aspect of religion—and scientific information.

The Pleasure of Finding Things Out

Turning to the third aspect of religion–the inspirational aspect–brings me to the central question that I would like to present to this imaginary panel. The source of inspiration today–for strength and for comfort–in any religion is very closely knit with the metaphysical aspect; that is, the inspiration comes from working for God, for obeying his will, feeling one with God. Emotional ties to the moral code–based in this manner–begin to be severely weakened when doubt, even a small amount of doubt, is expressed as to the existence of God; so when the belief in God becomes uncertain, this particular method of obtaining inspiration fails.

I don't know the answer to this central problem–the problem of maintaining the real value of religion, as a source of strength and of courage to most men, while, at the same time, not requiring an absolute faith in the metaphysical aspects.

The Heritages of Western Civilization

Western civilization, it seems to me, stands by two great heritages. One is the scientific spirit of adventure–the adventure into the unknown, an unknown which must be recognized as being unknown in order to be explored; the demand that the unanswerable mysteries of the universe remain unanswered; the attitude that all is uncertain; to summarize it–the humility of the intellect. The other great heritage is Christian ethics–the basis of action on love, the brotherhood of all men, the value of the individual–the humility of the spirit.

These two heritages are logically, thoroughly consistent. But logic is not all; one needs one's heart to follow an idea. If people are going back to religion, what are they going back to? Is the modern church a place to give comfort to a man who doubts God–more, one who disbelieves in God? Is the modern church a place to give comfort and encouragement to the value of such doubts? So far, have we not drawn strength and comfort to maintain the one or the other of these consistent heritages in a way which attacks the values of the other? Is this unavoidable? How can we draw inspiration to support these two pillars of Western civilization so that they may stand together in full vigor, mutually unafraid? Is this not the central problem of our time?

I put it up to the panel for discussion.

Permission Acknowlegdments

"The Pleasure of Finding Things Out" is the edited transcript of an interview with Richard P. Feynman that was broadcast as a BBC2 television program called "Horizon: The Pleasure of Finding Things Out." It is reprinted with permission of the producer Christopher Syckes, Carl Feynman, and Michelle Feynman.

"Computing Machines in the Future" was originally published in 1985 as a Nishina Memorial Lecture. It is reprinted here with kind permission of Professor K. Nishijima on behalf of the Nishina Memorial Foundation.

"Los Alamos from Below" was originally published by the California Institute of Technology in *Engineering and Science* magazine. It is reprinted with permission.

"What Is and What Should Be the Role of Scientific Culture in Modern Society" is reprinted with permission of the Societa Italiana di Fisica.

"There's Plenty of Room at the Bottom" was originally published by the California Institute of Technology in *Engineering and Science* magazine. It is reprinted with permission.

"The Value of Science" is from *What Do You Care What Other People Think?: Further Adventures of a Curious Character* by Richard P. Feynman as told to Ralph Leighton. Copyright © 1988 by Gweneth Feynman and Ralph Leighton. Reprinted with permission of W.W. Norton & Company, Inc.

The Pleasure of Finding Things Out

Index